Foreground Chemistry 3—*Rates of Reaction and Equilibria*

Editor's Foreword to the Series

Initial experience of teaching the Nuffield O-Level Chemistry Course, for a number of us involved in the Project, suggested that there was a need for more foreground material and more questions for homework. This opinion has been reinforced by further experience and by the comments of a great many teachers who have used the Nuffield material.

The *Foreground Chemistry* series is designed specifically to meet this need. At the same time we believe that the books will be of value for other modern chemistry courses, in particular the Scottish 'O' Grade Syllabus.

It is intended that the books should be read after the topic has been covered experimentally. Experimental details are therefore not given in the text. The books may also be used for revision and as a source of questions for homework. They might well solve the problem of the boy or girl who has missed important parts of the course through absence.

The authors chosen for the series have all had first hand experience of the Nuffield Chemistry Course, either as area leaders or as members of the headquarters team. They are thus well qualified to write about their subjects in the spirit which the Nuffield Course conveys. We hope that our readers will enjoy the books.

In response to demand this new edition has been produced using SI units. In so far as possible the recommendations of the Association for Science Education have been followed.

Martin Rogers

Foreground Chemistry

General Editor **Martin Rogers**

Headmaster, Malvern College

Questions prepared by **Derek Stebbens,**
Senior Chemistry Master,
Westminster School

The Equilibrists—they move
to keep in the same place
(*Radio Times Hulton*)

Foreground Chemistry 3

Rates of Reaction and Equilibria

by

Clifford Othen

Senior Lecturer in Education, University College, Cardiff

Second Edition in SI Units

Heinemann Educational Books Ltd
London

Heinemann Educational Books Ltd

LONDON EDINBURGH MELBOURNE TORONTO
AUCKLAND JOHANNESBURG SINGAPORE NAIROBI
IBADAN HONG KONG NEW DELHI KUALA LUMPUR

ISBN 0 435 64303 7

Published by Heinemann Educational Books Ltd
48 Charles Street, London W1X 8AH
Printed in Great Britain by
Butler & Tanner Ltd, Frome and London

Preface

Many of the chemical changes which you have investigated have taken place fairly quickly, even if sometimes heating has been necessary to bring about the reaction.

However, there are a great many chemical reactions which, under normal conditions, are either too slow or, more rarely, too fast for convenience. If we are to make use of such reactions—and some of them could be very important—we need to understand how they can be speeded up or slowed down, according to our needs. In the case of those reactions which normally stop after only a small proportion of the possible product has been formed, we need to know how to increase the yield of the desired material.

This little book discusses some of these topics and in so doing introduces you to ideas which are of the greatest importance when chemistry is applied to the large-scale manufacture of the many chemical substances which play such an essential part in our life today.

The material is presented largely in the form of a conversation which includes many questions. It is hoped that you will pause and frame your own answer to each question before reading any further. Sometimes the answer will be found in the text a few lines later on, but the answers to some of the other 'thinking' questions are suggested in the appendix.

My thanks are due to Anthony Othen, who took most of the photographs, and Malcolm G. Lee, who assisted with much of the practical work, particularly the construction of the equilibrium box.

C. W. O.

Note on the Questions

The approach to Chemistry teaching and learning which is represented by the Nuffield 'O' Level Course, and some other modern courses, is one which lays emphasis on the ability to understand the subject and to reason in relatively new chemical situations.

It is clear that any such fresh emphasis calls for a new kind of revision and examination question. This series of books makes special provision for many questions which, in the main, test understanding and application rather than factual repetition. Consequently it is important to the whole approach of these books, that the pupils who use them should make a regular and sustained attempt to think about and answer as many as possible of the questions which are included in the Appendix.

These questions have been graded into three categories. The first category should be attempted by every pupil, and most pupils should try the second. The third provides further problems for pupils who wish to pursue the subject in greater depth.

Contents

* References in parentheses are to the Nuffield O-level Chemistry Course.

Part Three. Further Study of Chemical Equilibria

Part One

Reversible Changes

1.1. *Looking at heating changes*

At an early stage in your study of chemistry you will almost certainly have investigated the effect of heating a variety of different substances.

Can you recall some chemicals which you have heated and the results which you obtained?

Perhaps you heated magnesium in air; if so, you will not have forgotten the dazzling light which you saw when the metal was burning. Do you remember the white powder that was left when the reaction was over? *What was the chemical name of this powder?* If you kept some of it, did you find that it returned to the bright metal with which you began the experiment? No, this did not happen, and because of this, such a change as the conversion of magnesium to magnesium(II) oxide is an irreversible change. This is a chemical change which is not reversed by a simple reversal of the conditions which brought it about; thus the heating of magnesium in air produces magnesium(II) oxide, but the cooling of magnesium(II) oxide does not lead to the reformation of magnesium.

$$\text{Magnesium} + \text{Oxygen (from the air)} \rightarrow \text{Magnesium(II) oxide}$$
$$\text{or} \qquad 2Mg(s) + O_2(g) \rightarrow 2MgO(s)$$

Do not think that what has been said means that it is impossible to convert magnesium(II) oxide into magnesium metal. In fact, this is done on a large scale in one of the processes for the manufacture of magnesium, but the conversion cannot be brought about by a simple reversal of the conditions which changed magnesium to its oxide.

In the Pidgeon process for the manufacture of magnesium, magnesium(II) oxide is obtained from a natural form of magnesium(II) carbonate called magnesite. The magnesium(II) oxide is then mixed with coke and strongly heated in an electric furnace connected to a powerful vacuum pump. Under these conditions the magnesium, which is produced in the furnace, distils over into a cooling system where the vapour condenses to give the metal.

You will certainly know many other chemical changes produced by heating which are also irreversible.

Recall, or try to find out, what happens when:
(a) Copper is heated in air
(b) Red lead oxide is strongly heated in an open crucible
Are these irreversible changes?

You may also have examined the effect of heating zinc(II) oxide. If so, you will remember that this chemical changes when it becomes hot; the white powder turns yellow. But as soon as you take the tube away from the flame and allow the zinc(II) oxide to cool, the white colour returns. If you were able to carry out a chemical analysis you would find that the white powder you started with, the yellow powder obtained on heating and the white powder left after cooling, are all three zinc(II) oxide and nothing else. Heating has changed one form of zinc(II) oxide into another, but on cooling the change is reversed and the white form is reproduced.

Such a change is understandably called a reversible change. We could sum up this knowledge by writing the two changes in this form:

$$\text{White zinc(II) oxide} \xrightleftharpoons[\text{cooling}]{\text{heating}} \text{Yellow zinc(II) oxide}$$

where ⇄ or as it is sometimes written ⇌ is the sign for a reversible reaction.

From the experiments which you may have carried out you can probably answer these questions: *Why is the gentle heating of roll sulphur a reversible change? Why is the heating of iodine crystals also a reversible change? What is unusual about the heating of iodine? What is the special name given to the direct cooling of a vapour to form a solid?*

Now recall another investigation. When dry powdered blue crystals of copper(II) sulphate were heated, a vapour was given off and a white residue remained. If the vapour was condensed it could be shown to be water, and the white residue, which was copper(II) sulphate crystals without their original water, was called anhydrous copper(II) sulphate.

Copper(II) sulphate crystals →
        Anhydrous copper(II) sulphate +
                Water (in the form of steam)

But an even more interesting part of the experiment

2

occurred when you added a few drops of water to the cold, white anhydrous copper(II) sulphate. Not only did you get back the blue colour with which you began, but you also recovered the heat energy which you had to supply to bring about the change. *How did you observe or detect this release of energy?* Thus we could write the statement for a second reaction:

Anhydrous copper(II) sulphate + Water →
               Crystalline copper(II) sulphate

We could combine both statements into one and write:

Copper(II) sulphate crystals ⇌
           Anhydrous copper(II) sulphate + Water

To sum up, we can say that some of the changes which occur when solids are heated are irreversible but others are reversible, though the distinction may not always be as simple as this makes it appear.

By a reversible change we mean one in which the products formed are capable of reacting with each other to reform the original substance or substances.

1.2. *A reversible change of great historical importance*

During your study of the chemistry of the air you may have learned of the part played by Joseph Priestley and Antoine Lavoisier in solving the problem of burning.

You saw that when the red powder, mercury(II) oxide, was strongly heated it decomposed into mercury and oxygen:

$$\text{Mercury(II) oxide} \rightarrow \text{Mercury} + \text{Oxygen}$$
$$2HgO(s) \quad \rightarrow \quad 2Hg(l) \ + \ O_2(g)$$

You also found that when mercury is carefully heated for several days in air, to a temperature of about 300–350 °C, it slowly combined with oxygen to form some mercury(II) oxide.

$$\text{Mercury} + \text{Oxygen (from the air)} \rightarrow \text{Mercury(II) oxide}$$

But if mercury(II) oxide is decomposed by heating, why is it that this compound does not split up again as soon as it is formed from the mercury heated in air? Part of the answer is in the words 'strongly' and 'carefully', used of the two heating actions; it is partly a question of temperature. The interaction of mercury and oxygen is another example of a reversible change, and it is one in which temperature is an important factor in influencing the direction of the change.

We can write both changes in this form:

$$2HgO(s) \underset{350\,°C}{\overset{800\,°C}{\rightleftharpoons}} 2Hg(l) + O_2(g)$$

It was in 1777 that Lavoisier carried out his famous experiment, in which he heated a weighed quantity of mercury for twelve days and obtained a red powder, from which he later collected and measured a volume of gas which we now call oxygen. It was this experiment which first established conclusively that the air contains at least two different gases and it played an important part in the working out of Lavoisier's oxygen theory of combustion.

1.3. *Other examples of reversible changes*

(*a*) You may have carried out an investigation of the mineral limestone. If so, you will know that it is one of the natural forms of calcium carbonate and that when it is strongly heated it decomposes into calcium oxide (commonly called lime or quicklime) and carbon dioxide.

What you may not know is that if carbon dioxide is passed over heated calcium oxide it is possible to show that some calcium carbonate is reformed.

Can you think of a possible way of carrying out this experiment?

Consider carefully the precautions that would be necessary before you could claim to have shown that some calcium carbonate had been produced as a result of the reaction which you had carried out. For example, you would need to satisfy yourself that there was no calcium carbonate present in the calcium oxide with which you began. How would you do this?

This reversible reaction can be expressed in the usual way:

$$CaCO_3(s) \rightleftharpoons CaO(s) + CO_2(g)$$

(*b*) During your study of the chemistry of water you may have passed steam over heated iron filings and obtained hydrogen. The other product of this reaction was an oxide of iron with the chemical formula Fe_3O_4 and the balanced equation for this reaction is

$$3Fe(s) + 4H_2O(l) \rightarrow Fe_3O_4(s) + 4H_2(g)$$

How could we attempt to find out whether the products of this reaction (the iron(II) diiron(III) oxide and the hydrogen) can reform the original reactants (iron and steam)? Well, you say, 'Pass hydrogen over some of the hot oxide.' Yes, but certain precautions will be essential. If a reverse reaction does occur we will only establish this if we can show, beyond doubt, that steam and iron are formed. Therefore it follows that we must first be absolutely certain that the two substances with which we are to begin do not contain any iron or water.

How would you ensure this? Think out the apparatus you would use, the precautions you would take and the tests you would apply to find out whether the reverse reaction had in fact occurred.

Careful practical work enables us to write

$$3Fe(s) + 4H_2O(l) \rightleftharpoons Fe_3O_4(s) + 4H_2(g)$$

Notice the use of the terms 'forward reaction' and 'backward reaction' or 'reverse reaction'. Of course if we had begun our investigation the other way round these terms would be interchangeable. Notice also the terms 'reactants' and 'products'.

1.4. *To help us with an important idea*

Before we consider this iron and steam reaction more fully, we need to get used to thinking of a rather special kind of balancing.

We will first imagine two fairly familiar situations in which we can see the principle at work.

(*a*) *At the circus*

Have you ever seen what is called 'a high wire act' at a circus? In this the performer balances on a steel wire high up in the circus tent. He often carries a long pole which stretches out on either side of his body, and if you watch very carefully as he stands 'still' in the centre of his walk, you can see that he is moving the long pole very gently.

Fig. 1

In what direction is he moving it?
Can you see why this movement helps to keep his balance?
If his body moves to one side, say his right, which end of the pole will he move and in which direction?

Perhaps you can now appreciate that his balance position is kept up by a series of small opposing movements. A small body movement in one direction is counteracted by a pole movement in the opposite direction. Such a circus performer is sometimes called an EQUILIBRIST. The special kind of

balance which 'keeps still by moving' is called EQUILIBRIUM.

Most of you will have done a low-level equilibrist act, when you walked on top of a wall keeping your balance by moving your outstretched hands up and down!

(b) At the Underground station—or the department store

In the photograph (Fig. 2) you see a pair of 'moving staircases' or escalators, one for carrying passengers, or shoppers, up to the next level and one for bringing them down from the floor above. Usually those being carried stand still on the step they occupy, but they are also asked to keep to one side of the staircase so that any who are in a hurry may, if they feel sufficiently energetic, run up the staircase which is already moving slowly upwards. In this case their speed upwards is the sum of two components, their own rate of running plus the upward velocity of the staircase.

But I have been at a London Underground station when the upward-moving staircase has been so crowded that a boy who was in a very great hurry decided to use the empty downward staircase as his route upwards. How did he do it? He ran very fast—thus using a great deal of energy—for he had to move upwards faster than the staircase was moving down. Only thus would he have a resultant upward velocity and thus ultimately reach the top.

If you look carefully at the photograph (Fig. 2) which has been specially taken in a department store, you will see that the young man on the left is running down the upward-moving staircase. (Notice the surprised expression on the face of the other man who is using the proper staircase for the downward journey.) The young man on the left ran fast enough to reach the bottom in spite of the upward movement of the staircase.

But what would happen if his speed downwards had been exactly the same as that of the staircase upwards? He would certainly have been moving, that is running, but would he have made any progress forward or backward?

This is clearly another situation in which a position of equilibrium could be reached, for there would be no resultant overall movement due to the balancing of two continuous but opposing movements.

Fig. 2

You now have a better understanding of how there can be continuous movement of some parts of a system and yet, when viewed as a whole, there need be no overall change.

Here is an interesting piece of apparatus which enables us to explore this idea a good deal further.

1.5. *The equilibrium box*

The photograph (Fig. 3) shows a glass-fronted box divided into two compartments by a wooden partition, the height of which can be varied at will. The base of each compartment is

connected to a source of compressed air, conveniently obtained from a vacuum cleaner or a compressor.

A large number—several hundred—of small, very light, expanded polystyrene balls are placed in section A (the left-hand compartment) and the air supply to that section switched on. What do you expect to happen?

If you try the experiment you will be able to see if your forecast was correct. Scientists are constantly doing this: thinking out what might happen in a particular situation and then trying it out practically to see if their conclusions are verified on the laboratory bench.

Wait until the change is complete and no further movement of the balls takes place. If the air supply to A is now cut off and the supply to B (the right-hand compartment) is switched on, what do you expect to happen?

Not surprisingly we finally come back to the original situation with all the balls in A. Finally—yes! But did you notice any difference between the time taken for the first few balls to

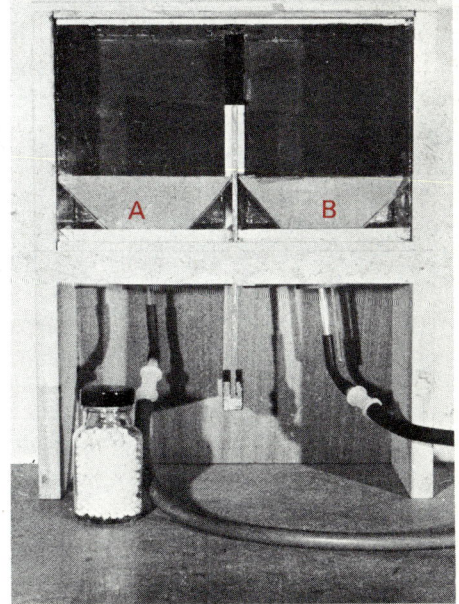

Fig. 3

pass over the barrier compared with that taken by the last few? We shall return to this important point later.

Now for another experiment.

The balls are all in compartment A. We switch on both air supplies so that an equal pressure of air is blown into each compartment and we watch carefully what happens. We are content to get a general overall impression of the situation. After a few seconds we see that most of the balls are still in compartment A though there are some in compartment B. *As the seconds pass, what do you think we notice about the number of balls in B? What must follow about the number in A?* After a few minutes have passed and we continue to examine the box, we seem to have the general impression that no further total change is taking place.

If a few balls are made distinctive by colouring them, then you can follow their position quite easily, and this will help you to appreciate what is happening.

If you switch off both air supplies simultaneously you can count the number of balls in each section when there appears to be no further visible total change. What, very roughly, do you expect to find? Did you, in fact, find this to be the case?

Can you now see clearly how it would be possible to have a situation of balanced movement in which there was, very nearly, the same total number of balls in each section? Yet all the while the individual balls, which make up the nearly constant totals, are still frequently changing. When this position, which we know as one of *dynamic equilibrium*, has been reached, then in the time that, say, fifty balls pass over from A to B, another fifty balls pass from B to A.

Could any of the second fifty be the same as some of the first fifty?

Now let us use a screw clip on the rubber tube delivering the air to section B, so that the air pressure supplied to A is much greater than that supplied to B. Place all the balls in section A and then switch on the air. After a few minutes when, as far as you can judge, no further total change is taking place, are there more balls in A or in B? Does this surprise you? How can it happen that although there appear to be (and we can

check that is the case by switching off the air and counting the balls in each compartment) many more balls in B than there are in A, yet the total position appears to remain unchanged? This can only be the case if, in the time that, say, fifty balls pass from A to B, then, on average, fifty pass back from B to A; that is, the rates of transfer of balls in both directions are the same. And this, remember, when the total number of balls in one section is very different from that in the other, but the total numbers in each section are not, on average, changing to any appreciable extent. Clearly how different these total equilibrium numbers are, will depend on how much greater the air pressure is on one side than on the other. By altering the relative air pressures we could arrange a whole variety of equilibrium positions. But in all cases once equilibrium has been established, then the opposing rates of ball transfer are EQUAL, however UNEQUAL but constant the different total numbers of balls in each section may be. It is

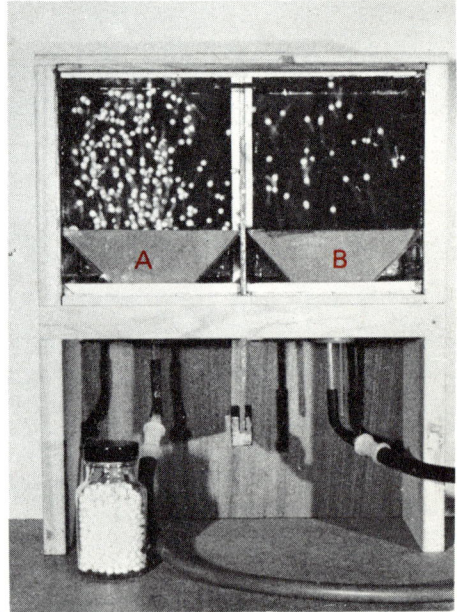

Fig. 4

6

very important that you should be quite clear that, apart from a slight variation of a few balls occasionally, when dynamic equilibrium is reached the opposing rates of transfer are equal, even though the numbers of balls in each compartment are unequal but separately constant.

The two photographs (Fig. 4 and Fig. 5) show two stages in the attainment of the equilibrium in one such experiment.

From our thinking about high-wire equilibrists, our misuse of a moving staircase and our experiments with the equilibrium box, we are equipped with an idea which has proved very useful in accounting for a situation which can arise in some chemical reactions.

In the case of a reversible reaction we have a possible two-way movement. Could this give rise to a position of dynamic equilibrium? This is our next question and we will now return to more detailed examination of the iron and steam reaction.

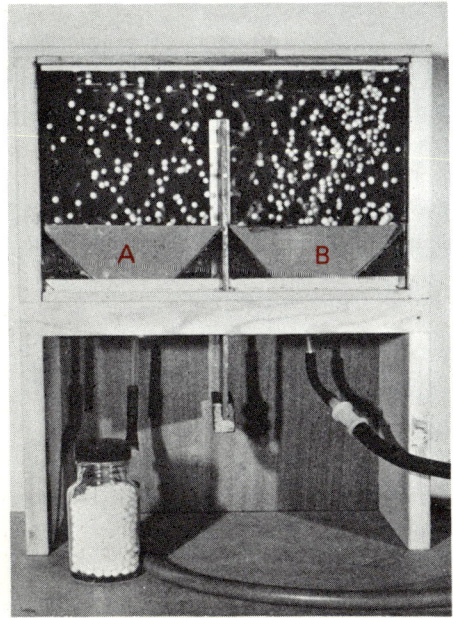

Fig. 5

1.6. *A more detailed consideration of a reversible reaction*

Let us now consider this last reaction more carefully.

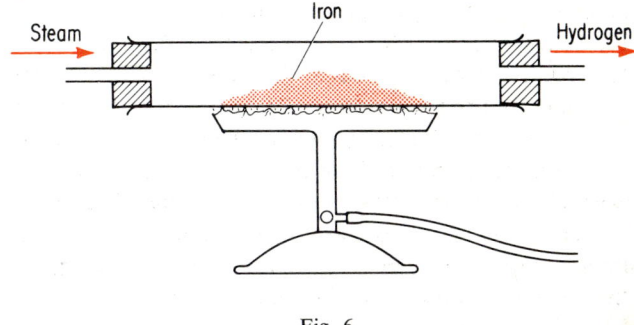

Fig. 6

When some of the steam which is passing through the tube meets hot particles of iron a chemical reaction may occur. If it does, some iron will be oxidized to iron(II) diiron(III) oxide and the reacting molecules of steam will be reduced to hydrogen. This hydrogen will be swept away in the current of steam and thus it will not be in a position to react with the iron(II) diiron(III) oxide to reform iron. Hence it should be possible to convert all the original iron into iron(II) diiron(III) oxide (though it might be difficult to achieve this) and the forward reaction is then said to have gone to completion.

But what would be the situation if the hydrogen was not removed as soon as it was formed?

Consider a heated, closed vessel containing iron and steam.

This is not an experiment which it would be easy or safe to attempt, but it is helpful to consider the situation.

At the outset the hot vessel contains only iron and steam. If the temperature is increased until the reaction is possible then, at some favourable points, steam molecules will react with iron atoms to form some iron(II) diiron(III) oxide and hydrogen. This small number of initial molecules of hydrogen will be mixed with a much greater number of steam molecules, and the chance of a reaction between hydrogen and iron(II) diiron(III) oxide, to reform the metal, will be very small (but

7

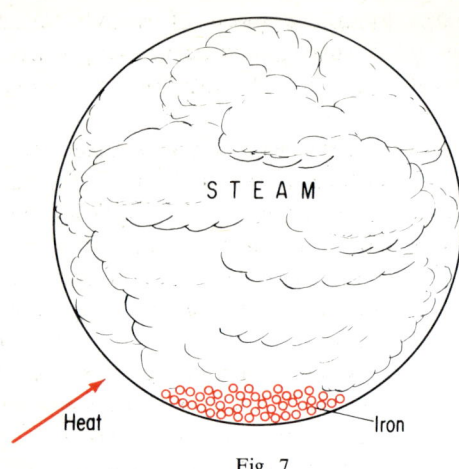

STEAM

Heat

Iron

Fig. 7

not zero) and the speed of this change will probably be so slow as to be almost negligible. But as the heating continues more iron(II) diiron(III) oxide and hydrogen will be formed from the iron and steam, and the opportunity for the reverse reaction to occur will be consequently greater and it would seem likely that the speed of this 'backward' reaction will increase steadily. (We shall consider ideas about speeds of chemical reactions more fully at a later stage.) In the meantime, with the removal of some of the iron, as it is converted into iron(II) diiron(III) oxide, and the subsequent diluting of the steam with the hydrogen which is formed, the possibility of steam molecules meeting iron atoms becomes much less and the speed of the forward reaction might be expected to decrease accordingly.

It is not difficult to envisage a point when a special kind of balance would be reached, so that over a short interval of time, some further small quantity of iron is converted into iron(II) diiron(III) oxide whilst at other points in the vessel a corresponding amount of iron(II) diiron(III) oxide will react with hydrogen to reform the same amount of iron. Thus whilst the total mass of iron (and also of the other three substances present) will not alter any further, yet the particular particles which constitute that constant total mass are continually

8

changing. There is no OBSERVABLE TOTAL change yet the situation is not unchanging, or static, for both the forward and the backward reactions are continuing at the same speed. Such a special condition of balance, when there is no further observable total change in the masses of the reactants and the products, yet when there is still continuous change at the molecular or atomic level, is known as DYNAMIC EQUILIBRIUM. This position of dynamic equilibrium can only occur when the forward reaction is proceeding at the same rate as the reverse reaction.

Can you see that the same reasoning could be applied if we had considered a situation in which our starting materials were hydrogen and iron(II) diiron(III) oxide? We should, in our theory, reach exactly the same point of dynamic equilibrium with an identical equilibrium state, assuming the temperature was the same.

Remember that the equilibrium position in this reaction could only be imagined to occur in a closed vessel where the products of the forward reaction are retained and thus make it possible for the reverse reaction to occur. The continuous removal of either product, as occurs when a current of steam is passed along a heated tube containing iron, would clearly prevent the reverse reaction taking place and hence make impossible the attainment of the equilibrium position.

Reversible reactions might thus, so we reason, in some circumstances, reach a position of dynamic equilibrium, whilst under other conditions they can be carried to completion and thus behave as though they were irreversible reactions (or nearly so). So far we have not produced any experimental evidence for this idea of dynamic equilibrium, but a scientist often argues from an idea to see where it leads and then he looks for experimental support for the conclusions which seem to follow from his original idea. In the coming pages we shall examine several of these ideas, the speed of a chemical reaction and the possible attainment of an equilibrium position, for they are very important in the study of chemistry.

But before we move on to do this, we need to have a more detailed mental picture of what might be happening at the atomic or molecular level when a chemical reaction occurs. As we construct our mental model we shall be meeting ideas which are capable of very advanced treatment, but we shall keep our thinking at a relatively simple, but nevertheless useful, level.

1.7. *Trying to picture what happens in a chemical reaction*

We all know that before a chemical reaction can take place between two substances, which are capable of reacting, they must be brought into contact.

You will have discussed earlier in your course the evidence for believing that all matter is made up of very small particles —atoms, molecules or ions—and that these particles are in varying degrees of motion according to whether they are in a solid, a liquid or a gas.

By bringing two substances into contact, as for example when we mix two liquids or two gases, we make it possible for the moving particles of each substance to bump into each other in a series of collisions. This picture of chemical reactions occurring as the result of collisions between the reacting particles has proved to be a very useful model, capable of both qualitative and quantitative interpretations.

According to this 'collision hypothesis' model, collisions between reacting particles are necessary before a reaction will occur. But then you may want to ask, 'Do all collisions result in chemical action?'

Think about the investigations you carried out into the rusting of iron. Picture a piece of freshly polished iron exposed to the air. If rusting is to occur, oxygen from the air has to react with the iron. (The total process is much more complex than this.)

It seems clear that very many oxygen molecules must be colliding with the surface atoms of the iron, yet no obvious reaction occurs in a short time. If every collision between an oxygen molecule and an iron atom resulted in a reaction, we would expect a film of red oxide (rust) to form immediately. As this does not happen it seems reasonable to conclude that not every collision results in a reaction, in fact it would appear that most collisions are ineffective chemically. But why is this? If only some collisions appear to lead to reaction, then all collisions are not alike in every respect. It is not difficult to make a guess as to the possible cause of this difference between one collision and another.

Imagine a closed flask containing oxygen at a constant temperature (Fig. 9). The molecules of the gas are in rapid movement.

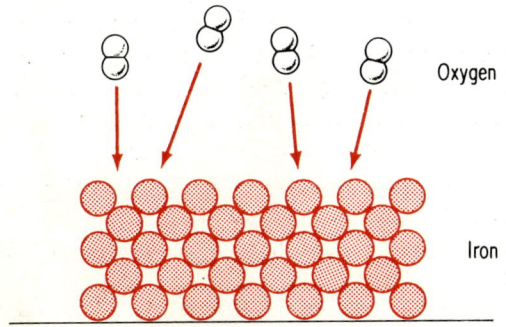

Fig. 8

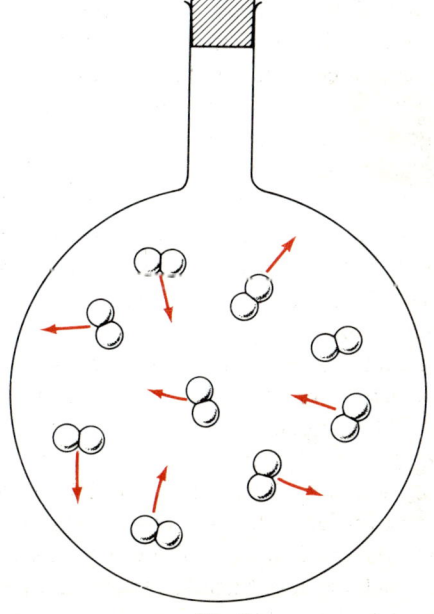

Fig. 9

9

Let us assume that at one particular moment all the vast number of molecules (27 million, million, million/cm³) are moving with the same velocity and thus all possess the same energy. An enormous number of molecular collisions takes place every second—this number has been calculated to be something like 10^{30} each second (that is, a million, million, million, million, million per second). These collisions are likely to occur somewhat unevenly. Thus a gas molecule at one spot in the flask might be subjected to more collisions than one at some other spot. Also the collisions are likely to vary in type; some might be head-on crashes (Fig. 10):

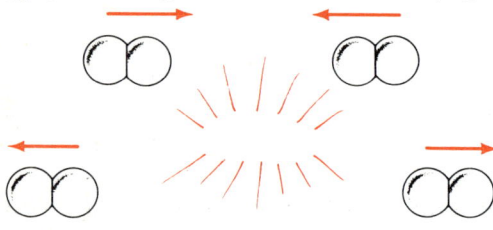

Fig. 10

others might be glancing blows, or spinning knocks (Fig. 11):

Fig. 11

Remember that the molecules as well as moving about from place to place are also, at the same time, rotating and vibrating.

Thus we can see that the number and nature, and hence the result, of the collisions suffered by individual molecules may vary greatly at different points in the gas. It seems likely that, in the short while from our imaginary starting point, those molecules which have suffered a certain sequence of col-

10

lisions may be moving much more rapidly than those which have experienced a different series of collisions.

Recall what happens with bumping cars on a fairground. At any given moment some are moving fast, others more slowly and some hardly at all. One may be spinning round without moving forward.

Those molecules which are moving, rotating or vibrating more rapidly than the average for the gas molecules in that flask, will clearly possess a higher energy than the average energy of all the molecules. A few molecules might possess a very much higher energy than the average.

It follows that some molecules are likely to possess lower energies and a few may possess very much lower energies. What we are picturing is a whole range of energies from well below, through average to above average and even well above average for a few.

You will be interested to know that it has been possible to carry out an experiment which confirms our picture. In this experiment a piece of metal was vaporized at one end of a vacuum tube and the vapour travelled along the tube to be deposited on a rotating disc at the other end. From the results it was possible to show that the metal atoms in the vapour had a range of energies. This energy spread is best represented by a graph (Fig. 12) which shows the fraction of the total number of atoms (in this case atoms of the metal) which have particular energies.

This shows that at a particular temperature a few atoms have a relatively low energy, the biggest fraction have about the average energy and different smaller fractions have considerably higher than average energies.

We now have a possible clue as to why all collisions between molecules capable of reacting do not, in fact, lead to a reaction.

It could be that only those molecules which have a certain energy, well above the average for that particular substance under its existing conditions, are sufficiently energetic to cause a collision which leads to a reaction. These highly energetic molecules may only be a very small proportion of the total and thus only a relatively small number of molecules

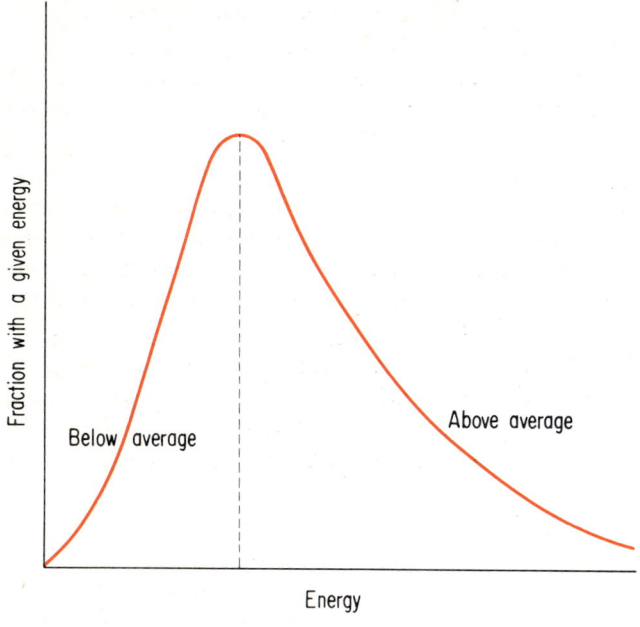

Fig. 12

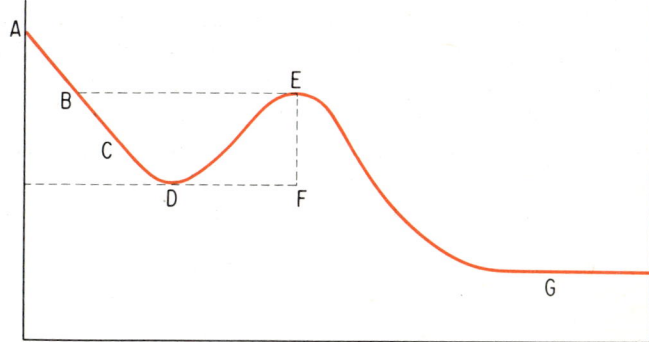

Fig. 13

would be capable of engaging in the reaction. This idea of a certain minimum energy—a kind of energy hill or barrier which must be surmounted before a reaction can occur—has proved most useful. Chemists speak of the ACTIVATION ENERGY of a particular reaction, by which they mean the minimum energy necessary to initiate the particular reaction being considered. This activation energy varies widely from reaction to reaction.

For some reactions it is so low that they occur immediately on first contact between the reactants. *Can you think of examples?*

In other cases external energy in the form of heat or light must be supplied before the activation energy barrier can be overcome and the reaction started. *Can you think of examples of the use of light energy to set off a reaction?*

We can be helped to imagine this idea of energy barriers by using different models.

Try rolling a large ball bearing down a track made of curtain rail or wood shaped like a hillock with uneven sides (Fig. 13).

If we hold the ball at C and then let go, will it pass over the hump at E? Where will it finally come to rest? If we release the ball at A where will it end up? What is the lowest point on the slope which will be just sufficient for the released ball to travel over the hump at E? What vertical height on the diagram might be said to correspond to the 'activation energy' of the reaction involved in crossing over?

You can have another mental picture of an energy barrier if you think back to our partitioned box with the barrier which can be varied in height (Fig. 14).

In which will the balls travel over into compartment B more quickly?

Or to put the same question in another way—in which case will less air pressure (that is, less energy) be needed to drive the balls from A to B in a given time?

You should now have a clear picture in your mind of the important idea of activation energy.

It often happens that if a chemical reaction can be started at one point then energy is released as a result of the action and this serves to activate other molecules nearby, so that by a kind of energy-release chain the reaction spreads throughout the whole of the mixture. Recall what happened

11

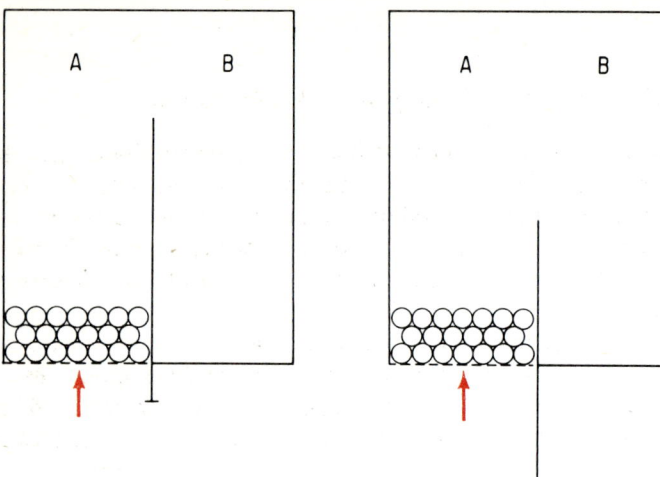

Fig. 14

There is another interesting possibility when we are considering what might happen in molecular collisions.

You have already learned something about the shape of molecules. You know, for example, how the chemist accounts for the differences between the properties of diamond and graphite in terms of the ways in which the carbon atoms are joined in these two substances.

If colliding molecules have shapes—or arrangements in space—might not these have some effect on a possible reaction? Think about possible collisions between hydrogen molecules and iodine molecules (Fig. 15):

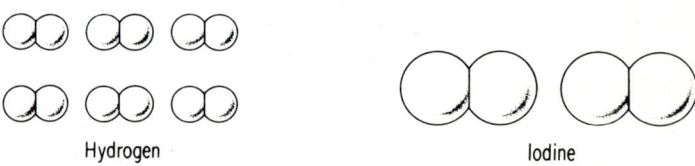

Hydrogen   Iodine

Fig. 15

when copper(II) oxide was heated in a stream of coal gas or hydrogen. Can you appreciate now why a spark from an electric wire or from a heat source can sometimes set off a chemical reaction?

We have tended to speak as though the energy necessary for a reaction was that contained in the movement (be it movement in space or rotation or vibration, or any combination of all three) of individual separate particles. In fact what is important is the joint energy supplied by two (very rarely more than two) colliding particles.

Think about a collision between two motor cars. Which is likely to cause the most damage? A head-on collision when they are moving in opposite directions, or a collision when one is stationary? The damage is really a measure of the energy released on collision. What about a glancing collision?
Similarly in molecular collisions the joint energy may be made up of two nearly equal contributions or of two very different contributions. But however it is reached, the total involved must exceed the activation energy for reaction to be possible between particular molecules.

Two of these molecules might collide end-on (Fig. 16):

Fig. 16

or they might collide at some angle to each other (Fig. 17):

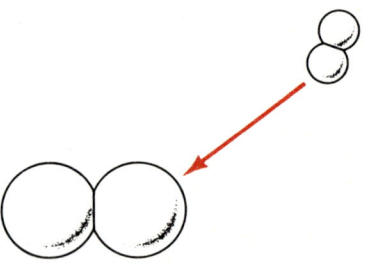

Fig. 17

12

or they might collide side-to-side (Fig. 18):

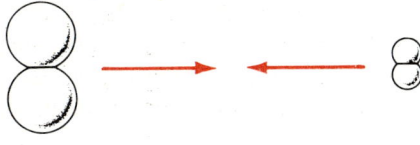

Fig. 18

Which collision position do you think is more likely to lead to the formation of two molecules of hydrogen iodide? But remember that even a side-to-side collision may only result in the two molecules bouncing apart, unless the joint energy of the colliding molecules is sufficient to exceed the activation energy of the reaction at the particular temperature of the two reactants.

We have been thinking of possible ways in which the molecules of iodine and hydrogen might collide. Until January 1967 chemists believed that the collisions which actually occurred when iodine and hydrogen reacted were of the side-to-side type between separate molecules of the two elements. Now an American research worker, John H. Sullivan, has carried out experiments which show that, before the reaction occurs, the iodine molecules are first broken into separate iodine atoms. Then two iodine atoms join on to a hydrogen molecule to make a new structure, from which the product molecules are finally formed.

A chemist must always be ready to change his views of the way in which a particular reaction takes place, when new experimental evidence is found.

1.8. *What really happens in molecular collisions?*

In the case of the reaction between iodine and hydrogen, the energy of collision must ultimately break two sets of bonds, those between the iodine atoms and those between the hydrogen atoms, and make new sets of bonds, those between the iodine and hydrogen atoms, to form the product, hydrogen iodide.

Speaking generally, in chemical reactions we are concerned with bond breaking and new bond making. In simple terms we might say that a chemical reaction takes the direction it does, because the bonds linking the reactants are more easily broken than those which join the different parts of the product.

If several bonds have to be broken and perhaps also major rearrangements made in the positions of the separated parts, then the energy necessary would be correspondingly greater— or what is the same thing, for a given energy supply, the reaction, if it went at all, would go slowly. Later you will understand why many reactions in what is called organic chemistry are usually rather slow.

You are seeing how helpful our molecular (or atomic) picture is, when coupled with our collision idea, in giving an account of several aspects of chemical reactions, so that the total picture fits together and makes sense. A scientist really cannot say whether his pictures (that is, his theories) are TRUE; he is more concerned that they seem to fit together and make a total picture which is useful because it suggests more ideas that he might try out.

Part Two

Rates of Reaction

Now that we have built up our molecular collision picture to help us understand how chemical reactions occur, we are ready to move on to learn how the speed of a reaction can be altered and hence controlled.

But first a few introductory words about reaction speeds, how they are defined and how they are measured.

2.1. *Different reactions proceed at different speeds*

It is time we moved back from our model to consider some chemical reactions and to think more about the speed at which they occur.

From the reactions which you have already studied it will be clear to you that there is a very great variation in the speed with which chemical changes take place.

Thus, if a solution of potassium iodide is added to one of lead nitrate, then as far as can be seen, the moment contact between the two solutions takes place, the reaction occurs and the yellow precipitate is immediately visible. Such a reaction would appear to be instantaneous. Other reactions, whilst not taking place instantaneously, nevertheless proceed at a very rapid rate; for example, if you light a length of magnesium ribbon or if you add a small piece of sodium to water, a very rapid chemical reaction will occur in each case. At the other extreme, a chemical reaction such as the rusting of iron or the formation of the green deposit on the surface of copper may take months or even years, according to the conditions, before there is any obvious sign of new products.

Thus we note that different chemical reactions proceed at very different rates, varying from those which are so rapid as to be violent or even explosive, to those which are so slow that it is difficult to detect any change. But all reactions proceed at a definite rate and many take place at rates which can be measured.

2.2. *What do we mean by the rate of a chemical reaction?*

Before attempting to measure the rate of some reactions we must be clear what we mean by the rate of a chemical reaction.

As a chemical reaction proceeds, reactants are being used up and hence disappearing, whilst products are being formed and thus appearing. Both the reactants and the products may be recognized by some suitably chosen property, as for instance colour or acidity (pH value). If we could measure the rate of either the disappearance of one of the reactants or the appearance of one of the products, we would be measuring the rate at which the reaction is proceeding. Thus we may be able to determine the rate of a reaction by selecting some suitable physical or chemical measure of the amount of either one of the reactants or of one of the products present, and then observing how this measure changes in a given period of time.

Thus, in general terms, we might define the rate of a chemical reaction:

$$\text{Rate of reaction} = \frac{\text{Measured change in a selected property}}{\text{Time taken for the change to occur}}$$

To be exact, this is the average rate of the reaction over the measured interval of time. If, however, the interval of time is very small, then we may speak of the rate of the reaction at that particular moment.

Consider an example, first from the point of the disappearance of a reactant.

If we added 96 g of magnesium to a measured quantity of sulphuric acid and allowed the reaction to proceed for exactly five minutes, and then found that there were exactly 24 g of magnesium left, we could say:

$$\text{Average rate of reaction} = \frac{96 - 24}{5} \text{ g of magnesium per minute}$$

where the average rate of the reaction is taken to be the average rate of the disappearance of magnesium during the time of the experiment.

It is more usual to express the rate in mol/sec.
Can you calculate this when you know that Mg = 24?
If you can you will find that it comes to 1/100 mol magnesium/second.

It would be equally possible, and perhaps more convenient, to measure the rate of the reaction in terms of the volume of hydrogen produced, either in cm^3/sec or in mol/sec.

Many different methods are employed to follow the progress of chemical reactions and thus to measure their rates. Changes in colour, in weight, in volume, in pressure, and in many other properties have been used. It is a question of which is most convenient for a particular reaction.

For fast reactions the unit of time may need to be a fraction of a second, such as a millisecond (1/1000 of a second) or even in some cases a microsecond or a nanosecond.
Can you find out what these are?

2.3. *Measuring the rate of a reaction in the laboratory*

It is time we tried to measure the rate of some suitable chemical reaction.

You are familiar with the action of marble chippings with hydrochloric acid leading to the formation of carbon dioxide.

$$CaCO_3(s) + 2H^+(aq) + 2Cl^-(aq) \rightarrow$$
$$Ca^{2+}(aq) + 2Cl^-(aq) + CO_2(g) + H_2O(l)$$

If we started with a known total mass of marble and acid, and allowed the reaction to proceed in an open-necked flask, the mouth of which contained a loose plug of cotton wool to prevent any escape of liquid spray, we would expect to find a loss in mass as the reaction proceeded. The loss in mass would indicate the mass of carbon dioxide given off in any particular time.

The experiment can be conveniently carried out by putting 40 cm^3 of 2 M hydrochloric acid in a 100 cm^3 conical flask, adding 20 g marble chips, placing a plug of cotton wool in the neck of the flask and then weighing the whole apparatus on a direct reading balance (preferably of the open top-pan type). If the flask with the acid and the marble chips have been weighed separately before they are mixed, the operation can be carried out more rapidly. At the moment at which the mass is recorded, start a stop-watch. Note the reading on the balance at minute intervals and continue for ten to fifteen minutes until there is no further change in mass.

So that we may try to obtain a clearer picture of what is happening as the reaction proceeds, we usually draw a graph of the two quantities which we have measured. In this case we know the various masses of carbon dioxide which have been produced by the interaction of the acid and the marble at certain definite times. If you have carried out this experiment you will know that the graph which you obtain from your results is similar to that drawn in Fig. 19.

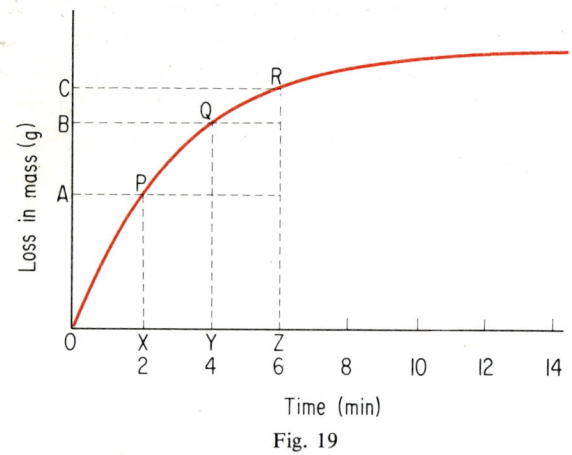

Fig. 19

Let us see what we can learn by studying this graph.

During the first two minutes of the reaction the mass of carbon dioxide given off is indicated by OA, which we will call $2w_1$ g.

During the second two minutes of the reaction the mass of carbon dioxide given off is indicated by AB, which we will call $2w_2$ g.

During the third two minutes of the reaction the mass of carbon dioxide given off is indicated by BC, which we will call $2w_3$ g.

From the shape of the curve it is clear that OA is greater than AB and that AB is greater than BC. In other words, the

16

amount of gas given off in each successive period of two minutes is getting less.

We can say that:
Average rate of the reaction during time OX
$$= \frac{2w_1}{2} \text{g of gas/minute}$$
$$= w_1 \text{ g of gas/minute}$$

Similarly
Average rate of the reaction during time XY
$$= w_2 \text{ g of gas/minute}$$
and
Average rate of the reaction during time YZ
$$= w_3 \text{ g of gas/minute}$$
This means that the rate of the reaction is decreasing as the time passes. [Expressing these rates in mol/s will not alter the conclusions.]

But our graph flattens out and finally becomes horizontal. *What is the rate of the reaction at this stage?*

This situation corresponds to the stage of the experiment when, although there were several pieces of unused marble in the flask, no further action was occurring. This would seem to suggest that there was no more acid left to react with the marble.

You could easily check this suggestion. How?

What picture do we now have of the progress of the reaction? At the start, with plenty of marble and fresh acid, the rate is most rapid. As the reaction continues, the rate slows down. Finally the reaction stops, although there is plenty of marble, because all the acid has been used up. It looks very much as if the reaction rate decreases as the acid is used up, so that we might write, for this reaction:

Highest concentration of acid . . . Fastest rate of reaction
Lowest concentration of acid . . . Lowest rate of reaction

For this reaction we can see that the concentration of the acid directly influences the rate of the reaction. This relationship is usually stated:

The rate of the reaction is directly proportional to the concentration of the acid when the marble is in excess.

You will notice that if we consider the separate sections of the graph OP, PQ, QR, as though they were straight lines, we should have (Fig. 20):

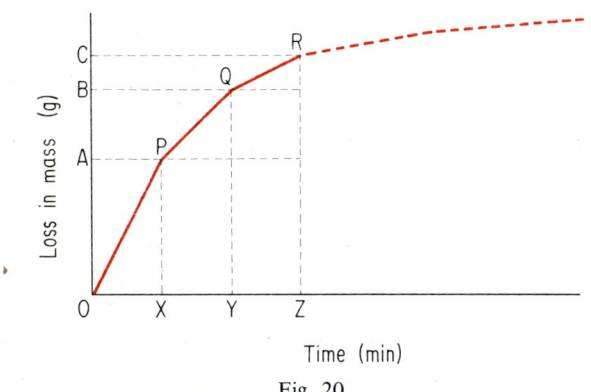

Fig. 20

If we think of these as hills or inclines, which is the steepest one?

We often speak of the slope or gradient of a curve at a particular point, and you will see that where the gradient is steepest the rate of reaction is fastest. The gradient of the line OP is measured by dividing PX by OX. We have already seen this is the measure of the average rate of the reaction during the time interval OX. In general, the gradient of the graph at any particular point measures the rate of the reaction at the time represented by that point.

2.4. *Investigating some of the factors that influence the rate of a reaction*

Now that you have learned how to measure the rate of this particular reaction, you may be able to carry out a series of experiments to find out how reaction rate is influenced by different factors.

For example, do you think that this reaction would proceed more slowly or more rapidly if you used the same concentra- tion of hydrochloric acid but added 20 g of crushed marble instead of chippings?

It may be possible for you to carry out this new experiment and from your graph obtain a value of the rate of the reaction at a point, say, four minutes after the start. You can then compare this value with the rate at the corresponding point in the first experiment.

Was your 'hunch' confirmed?

Or again, what effect do you think a change of acid concentration would have on the reaction rate for a fixed size of marble pieces? You could use 40 cm³ of 1 M hydrochloric acid with 20 g of the original size marble chips. Would you expect the reaction rate to increase or to decrease?

The two graphs in Fig. 21 show the rate curves obtained using excess of the same size marble chips, in the one case with 2 M acid and in the other with 1 M acid. *Can you say which is which?*

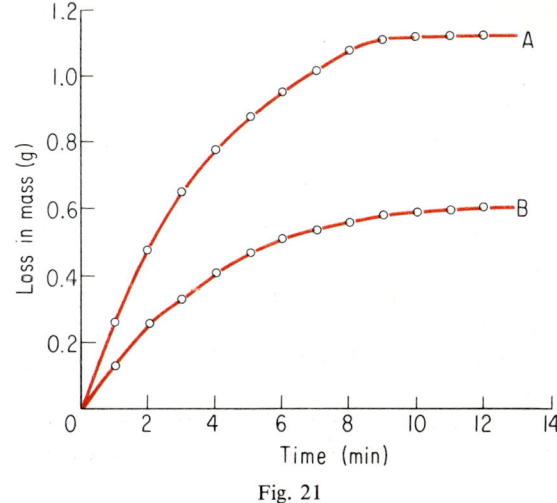

Fig. 21

Perhaps as a class you can carry out a wider range of experiments with different sizes of marble and different concentrations of acid. If you are able to examine such a set of

17

results you may feel able to venture an opinion, that for this reaction you could complete these two summaries:

(a) Decrease in particle size leads to . . . in the rate of the reaction, other conditions being the same.

(b) Increase in concentration leads to an . . . in the rate of the reaction, other conditions being the same.

2.5. *Another experiment to examine the effect of concentration on the rate of a reaction*

If you add some dilute hydrochloric acid to half a test-tube of sodium thiosulphate solution it will not be long before you see evidence of a chemical reaction.

You may have tried this already, in which case you will know that a yellowish precipitate of sulphur is soon formed.

The first appearance of a cloudiness, which is more readily observed by holding the test-tube over a piece of white paper, serves as a convenient way of telling when a certain fixed point has been reached in the reaction. Using a stop-watch, the time taken for this standard cloudiness to appear can be easily measured.

It is simple to vary the concentration of a given stock solution of sodium thiosulphate so that we obtain solutions which are 4/5, 3/5, 2/5, 1/5 or any other fraction of the original concentration. ***Can you work out how this is done?***

If we carry out a series of experiments with such solutions of varied concentrations, each time adding the same volume of the same hydrochloric acid, the cloudiness times which we measure will enable us to find out more about the effect of concentration on the rate of this reaction.

Here are some results obtained in this way (Fig. 22).

In each case the temperature was the same (room temperature) and the same volume of 2 M hydrochloric acid (5 cm^3) was added in each experiment.

From these figures we see that, as each solution gets weaker (because we dilute it with water), the time necessary to reach the same point of cloudiness becomes greater.

We would like to know whether there is any more definite connection between the various times recorded and the different concentrations of solution used. To examine this possible relationship we draw some graphs.

First we will plot a graph of concentration (which we can represent by the figure of 'the volume of stock solution used') against time.

Do you see why the volume of stock solution used in each experiment is a measure of the concentration of the sodium thiosulphate solution?

The graph we obtain is shown in Fig. 23.

Once again we have the same general pattern as in our experiments with calcium carbonate, but this time the curve is, as it were, the other way round. We begin, as before, with a fairly steep slope which gradually becomes less steep before it finally flattens out towards the horizontal. As we saw in the previous experiment, the gradient of the curve at any point measures the rate of the reaction at the time represented by that point on the graph. Once again we see that the rate of the reaction decreases as the concentration of the sodium thiosulphate solution decreases.

But it might be more interesting to try to plot a measure of the rate of the reaction against the concentration producing that particular rate. How can we do this easily?

Think about it in this way:

When the time taken for a given reaction is short, clearly the rate of the reaction is high. In the same way, when the time taken is long, then the rate is slow. In other words the rate is proportional to 1/TIME, which is called the reciprocal of the time; 1/T where T is the time.

The graph shown in Fig. 24 has been drawn by plotting the

Volume of stock solution	Volume of water added	Ratio of concentration of solution to that of stock solution	Time (sec)	1/Time
50	0	1	27	0.037
40	10	4/5	35	0.029
30	20	3/5	48	0.021
20	30	2/5	80	0.013
10	40	1/5	204	0.005

Fig. 22

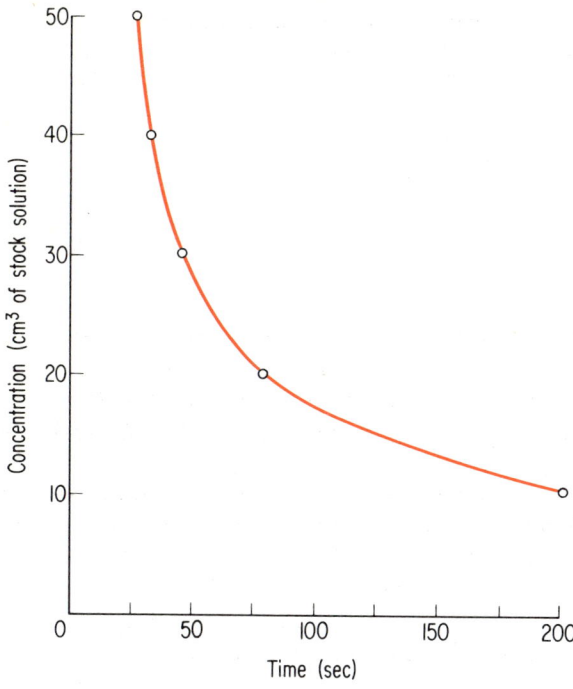

Fig. 23

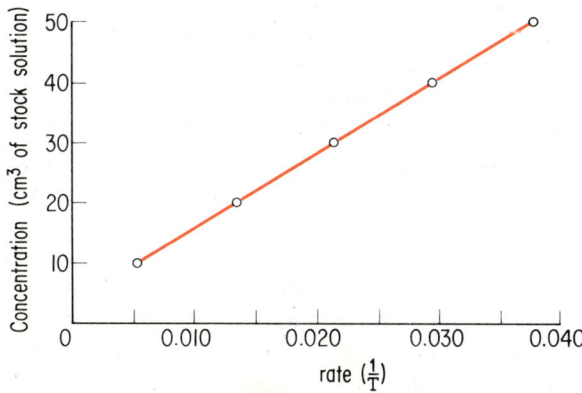

Fig. 24

concentration of sodium thiosulphate solution against the reciprocal of the time; the latter being a measure of the rate of the reaction for the appropriate concentration.

You notice that, within the limits of experimental error, the graph is a straight line. This represents a relationship which we express in words by saying that the rate of this reaction, at a fixed temperature, is directly proportional to the concentration of the sodium thiosulphate solution; provided, of course, that we use enough acid to react with all the sodium thiosulphate.

2.6. *How do changes in temperature affect the rate of a reaction?*

You are very familiar with the general effect of temperature on many chemical reactions. For example, you may be investigating the possible action of a certain metal with sulphuric acid. You add some cold acid to a piece of the metal and nothing appears to happen. Almost without thinking you gently warm the tube containing the reactants and you look again for signs that a reaction is occurring, now that you have supplied extra energy. In many cases the reaction, which was almost too slow to be seen at the lower temperature, is now readily apparent as it takes place much more rapidly.

We can use the reaction between hydrochloric acid and sodium thiosulphate to make some measurements from which the influence of temperature on the rate of a reaction will be very clearly shown. It is only necessary to use, for each experiment, a fixed quantity of hydrochloric acid and sodium thiosulphate, but before mixing the two solutions they are warmed to different temperatures each time and, of course, we record the temperature when the cloudiness times are reached.

The table in Fig. 25 shows the times taken for an agreed precipitate of sulphur to appear at different reaction temperatures.

You will at once notice two important points:

(*a*) At higher temperatures the reaction goes more quickly.

19

Temperature of reaction mixture (°C)	Cloudiness time (sec)
12	80
18	60
25	40
34	28
48	16

Fig. 25

You expected this, but you may not have anticipated the next point:

(b) A fairly small rise in temperature has a large influence on the reaction rate.

In this case a rise of temperature of 13 K (from 12 °C to 25 °C) doubles the speed of the reaction.

For many reactions it is approximately true that a rise in temperature of 10 °C leads to a doubling of the reaction rate. Remember that this will continue to happen for the next rise of 10 °C, and for the next! Can you see now why boiling a solution may lead to an enormous increase in reaction rate?

2.7. *Another way of influencing the rate of a reaction*

You may have seen your teacher carry out an experiment in which a comparison was made between the effect of heating potassium chlorate on its own (tube A) and the heating of a mixture of potassium chlorate with a little dry copper(II) oxide (tube B).

To make a fair comparison each tube received the same amount of heat and equally glowing splints were held expectantly at the mouths of the two tubes (Fig. 26).

(Can you suggest another simple way of heating the two tubes equally? *What gas did you hope to detect? Which splint rekindled first?*

In this way you can easily show that a mixture of potassium chlorate and copper(II) oxide decomposes to give off oxygen at a lower temperature than potassium chlorate heated alone. *How can we account for this?*

20

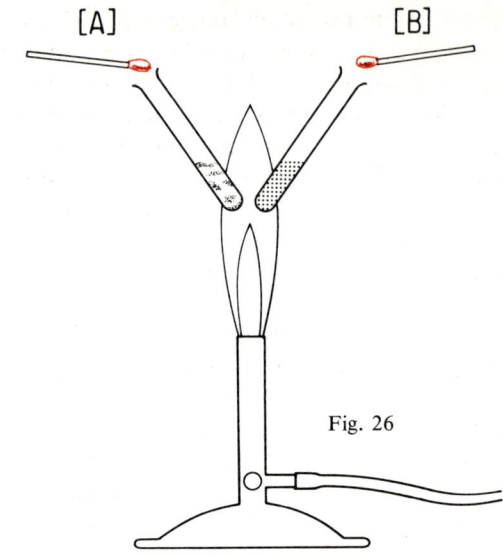

Fig. 26

Someone might suggest that the oxygen obtained at the lower temperature came from the decomposition of the copper(II) oxide. If this was true, what would you expect to find on heating dry copper(II) oxide alone? *How would you show that this does NOT happen?*

If you know that copper(II) oxide is insoluble in wa :r but that potassium chloride (left after the decomposition of the potassium chlorate) is soluble in water, you can devise an experiment to find out whether the copper(II) oxide is used up when it accelerates the evolution of oxygen from the potassium chlorate.

(*Hint:* Why not measure out a mass of 2 g of potassium chlorate and 0.5 g of copper(II) oxide. What next?)

As a result of experiments such as these you can show that the copper(II) oxide is not the source of the oxygen, that it is not consumed in the reaction but that, by its presence, it lowers the thermal decomposition temperature of potassium chlorate. Another way of describing the action of the copper(II) oxide is to say that it causes the decomposition to proceed more rapidly at a given temperature above the decomposition point.

You probably know that a substance which alters the rate of a chemical reaction, without itself being used up, is called a CATALYST. Notice that we said 'alters the rate'; why do you think we sometimes speak of positive catalysts and negative catalysts? Negative catalysts are more often called inhibitors, while positive catalysts are usually called just—catalysts.

Perhaps you have carried out experiments to show that several other chemical compounds—such as zinc(II) oxide and iron(III) oxide—also act as catalysts for the thermal decomposition of potassium chlorate.

Safety Warning. There have been some serious accidents arising from the heating of potassium chlorate and hence you need to exercise great care whenever this compound is heated either alone or mixed with other substances. Only use very small quantities and always heat in a small uncorked tube. Do not use this reaction as a laboratory preparation for oxygen.

2.8. *Another investigation into catalysis*

It is possible to carry out several similar investigations into the catalytic effect of various substances on other chemical reactions.

Here is one you may perhaps be able to try.

If you take a small quantity of 20 volume hydrogen peroxide in a test-tube you will not see any signs of a gas bubbling from it. If you warm the tube gently, bubbles will appear and you can easily identify the gas given off as oxygen.

You could now take a series of test-tubes containing a few cm^3 of hydrogen peroxide and try adding small amounts of different substances to each tube. You might choose substances from this list: copper(II) oxide, sodium chloride, manganese(IV) oxide, zinc(II) oxide, powdered charcoal.

It is soon clear that some of these substances make excellent catalysts for the decomposition of hydrogen peroxide at room temperature.

We are now able to use this knowledge to make a more careful study of a catalysed reaction.

2.9. *Investigating the rate of a catalysed reaction*

You have probably carried out several experiments using the reaction by which hydrogen peroxide is decomposed in the presence of a catalyst such as manganese(IV) oxide. It is easy to collect the oxygen evolved, and the use of a glass syringe simplifies the volume measurement.

If you use apparatus set up as shown in Fig. 27 it is possible to find out many things about the decomposition rate.

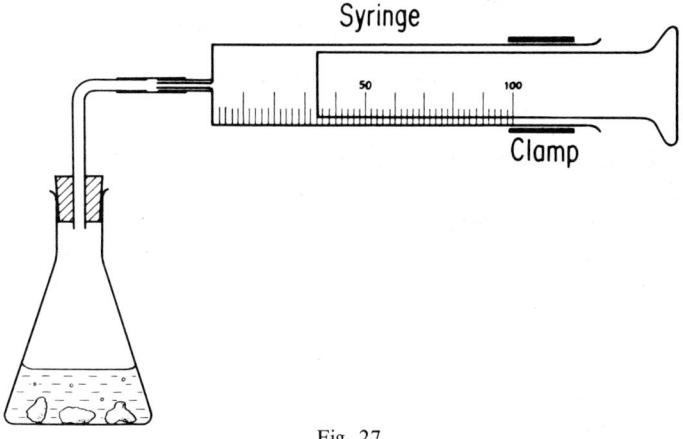

Fig. 27

Can you suggest some investigations which you might try if you have the opportunity?

Your first suggestion would probably be to find out what happens to the reaction rate when the concentration of the hydrogen peroxide is varied. This can be easily done by making up the reaction solution from stock 20 volume hydrogen peroxide, which has been diluted by different amounts.

Again you might wish to investigate the effect of temperature on the rate of the reaction whilst you keep other factors such as concentration, nature, and quantity of catalyst, the same. This also is easy to do by warming the reaction vessel before the catalyst is added.

You have already made similar investigations with other reactions, but it is important to get as much evidence as possible before we attempt to frame some general conclusions.

As well as confirming the results obtained in other experiments this reaction can also be used to find out new information. For example, we may like to know:

(a) Does the quantity of catalyst affect the rate of the reaction to any appreciable extent?

(b) How much better is one catalyst than another?

(c) Does it make a difference to the reaction rate if the catalyst is in granular form or in powdered form?

Perhaps by working in pairs on different questions such as these, your class has found out many things about the influence of different factors on the rate of this reaction.

Here are a few rate graphs for this reaction, obtained under various stated conditions (Fig. 28). Try to examine them carefully and make sure you understand what they mean.

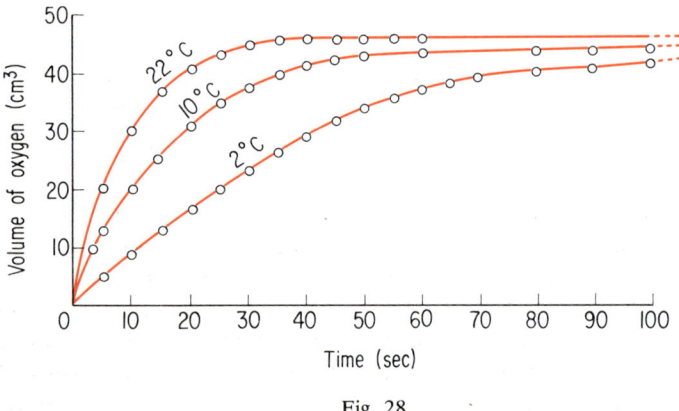

Fig. 28

2.10. *Summarizing our experimental results*

It is now time to sum up what you have learned about reaction rates from the various experiments which you have carried out or have seen demonstrated.

1. DIFFERENT CHEMICAL REACTIONS occur at different rates under identical conditions.

2. THE SAME CHEMICAL REACTION may occur at different rates depending on:

(a) The physical state of the reactants.

(b) The concentration of the reactants.

(c) The temperature of the reaction.

(d) The presence of a catalyst.

2.11. *Using our ideas to account for what we know*

By using the collision hypothesis and the idea of activation energy we have constructed a simple picture or model, which helps us to understand why reactions occur and why the rates of different reactions are not the same. Does this same model help us to make sense of the influence on reaction rate of physical state, concentration, temperature and catalysts? Let us consider them one at a time.

(a) *Physical state*

If a solid is in the form of a fine powder rather than large lumps, it is easy to see why we would expect a faster reaction rate.

Most reactions occur as a result of surface collisions. ***What can you say about the surface area of 1 g of powdered calcium carbonate compared with that of 1 g of lump calcium carbonate?***

What follows about the likely number of collisions in a given interval of time and hence about the reaction rate (Fig. 29).

(b) *Concentration of reactants*

When we use a more concentrated reagent instead of an equal volume of a more dilute one we are taking more particles of reactant into the same reacting volume. We are crowding the particles closer together and thus decreasing the distance they are likely to travel without a collision. (Think of the busy time on Saturday evening at the fun-fair bumping cars compared with the slack time on Monday afternoon. The space in which you can operate is the same, but if all the cars are out it will be more difficult to avoid a collision—assuming

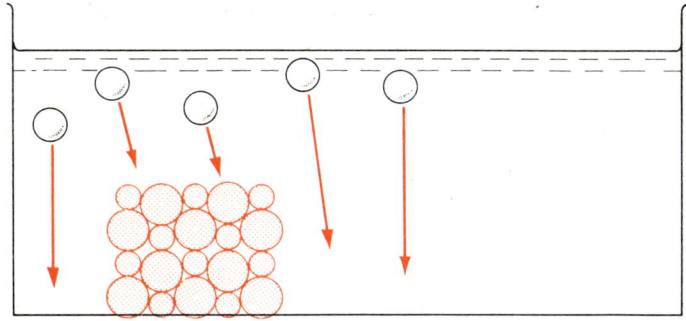

Solid in lump form

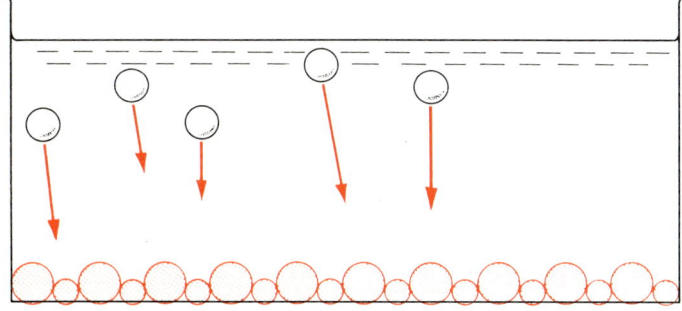

Powdered solid

Fig. 29

you wish to!) Thus, increasing the concentration reduces the length of the likely collision-free path and hence more frequent collisions occur (Fig. 30).

Again our picture would lead us to expect what you have already measured, an increase in reaction rate following an increase in the concentration of one or both reactants.

(c) Temperature

Here again the picture seems to fit easily. At a higher temperature we think of particles moving faster, and as a result collisions would be more frequent and reaction rate faster. (If the mechanic in the fairground suddenly speeded up every bumping car, then the time taken between collisions would decrease and the frequency of collisions would increase.)

But more advanced work shows that this is not the whole picture—or even the main part of it! It is possible to calculate what increased frequency of collisions would result from a speeding up of the particles, and this proves not to be sufficient to account for the very great increase in reaction rate produced by a relatively small rise in temperature. (Remember the rate of some reactions is doubled by a rise of 10 K.)

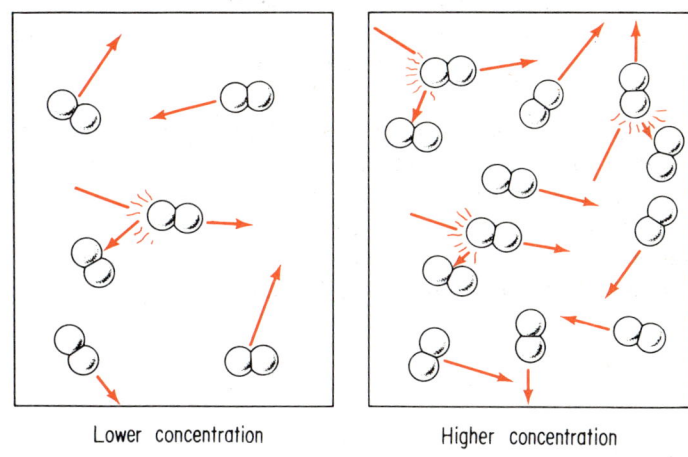

Lower concentration   Higher concentration

Fig. 30

We now know, what it would have been difficult to guess from our simple model, that the main effect of a temperature rise is to produce a rapid increase in the proportion of molecules with sufficiently high energies to make reaction possible. The graph in Fig. 31 shows that with rise of temperature there is a much greater proportional increase in the fraction of molecules with a higher-than-average energy. Compared with this effect, the increase in particle velocity is only a small factor—but it is a factor.

(d) Catalyst

Neither the effect of catalysis in general, nor any suggestion as to why one particular substance and not another acts as a catalyst for a given reaction, could be guessed at from our model.

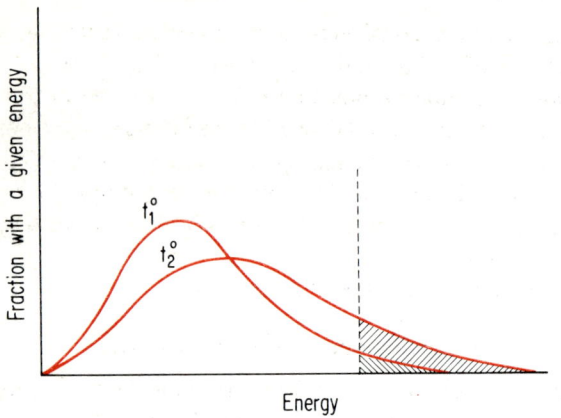

Fig. 31

Chemists now believe that a catalyst enables a reaction to proceed by an easier path, that is one requiring less energy to travel along it.

This picture may help (Fig. 32).

You might think of the energy required to cross the mountain range from A to B by climbing over the highest peak, compared with the energy required to reach B by using a tunnel cut halfway up the mountain.

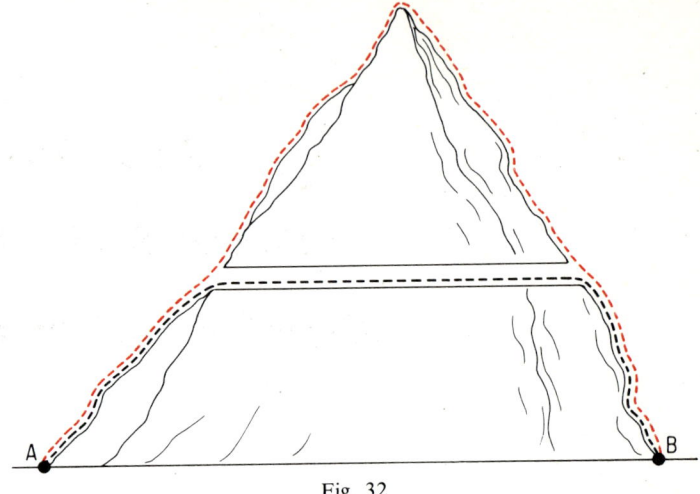

Fig. 32

Which path would require the expenditure of least energy on the part of the walker?

There is much more to the question of catalysis than we can discuss here. It is a very wide subject with tremendously important industrial applications. Try to read all you can, to find out how very greatly chemical industry depends on catalysts.

Part Three

Further Study of

Chemical Equilibria

3.1. *Looking at another reversible reaction*

You have already considered some chemical reactions which are capable of going both ways; we called them reversible reactions. It is time to look more closely at some of this type of reaction and thus extend our ideas of dynamic equilibrium.

Perhaps you have carried out an experiment in which you carefully added some sodium hydroxide solution to a weak solution of bromine water. If so you will know that the liquid becomes colourless.

We might represent the change thus:

$$\text{Yellow colour} \xrightarrow{\text{alkali}} \text{Colourless}$$

If this reaction is a reversible one, what might be done to restore the yellow colour? This really amounts to asking how you would attempt to counteract the effect of an alkali. You probably reason that the addition of a dilute acid would reverse the change and the indication that this had happened would be the return of the yellow colour.

A simple experiment shows that this reasoning is sound.

We can therefore now write:

$$\text{Yellow colour} \xleftarrow{\text{acid}} \text{Colourless}$$

But if this change can be made to happen once, why not a second and a third time? You probably repeated both changes, using alkali and acid alternately. Each time you found that the bromine water became colourless when sufficient alkali had been added and then the pale yellow colour returned with the careful addition of acid.

We have thus established the existence of a reversible reaction.

Thus we may write:

$$\text{Yellow coloured bromine water} \underset{\text{alkali}}{\overset{\text{acid}}{\rightleftarrows}} \text{Colourless solution}$$

The full chemical equation for the changes involved is

$$\underset{\text{Yellow}}{Br_2(aq)} + H_2O(1) \rightleftharpoons \underset{\text{Colourless}}{H^+(aq) + Br^-(aq) + HOBr(aq)}$$

We will now consider this in some detail.

25

The yellow colour, shown on the left-hand side of the equation, is caused by the dissolved bromine, provided, of course, that there is sufficient present for our eyes to detect it. All the other ions and molecules present are colourless.

When we add a few drops of bromine to water, a reaction occurs and some hydrogen ions, $H^+(aq)$, some bromide ions, $Br^-(aq)$, and some hypobromous acid molecules, HOBr, are formed. But as the reaction is reversible an equilibrium position is immediately reached when the rates of the two opposing reactions exactly balance.

Now consider what happens if we attempt to upset that balance.

If we add some alkali, as you will learn later, we are really removing hydrogen ions. This means that momentarily there will be insufficient to maintain the equilibrium. How can the balance be restored? Clearly a fresh supply of hydrogen ions is needed and these can only come from more bromine molecules reacting with more water molecules to form, among other products, some more hydrogen ions. This will happen and the change will continue until the equilibrium is once again established. But as we use up the source of the yellow colour—the bromine molecules—the solution will naturally become paler. If we add sufficient alkali, then the solution will lose its colour entirely as more and more bromine molecules change over to bromide ions, hydrogen ions and hypobromous acid molecules.

It should now be easy for you to understand how the opposite effect is produced; that is, why the addition of acid, which means the addition of hydrogen ions, causes the yellow colour to return. In this case the balance is being pushed back to the right-hand side of the equation and the reformed bromine makes itself visible in the restored colour.

3.2. *Two more similar reactions*

The previous investigation into the colour changes of bromine water in acid and alkaline solution probably made you think of the changes produced with indicators, such as litmus, when acid or alkali is added to them.

You may find it helpful to look again at these changes and try to account for them in terms of the upsetting and the restoring of an equilibrium.

It may be possible for you to carry out a similar reaction to the bromine water experiment, but this time using a solution of potassium chromate to which you first add a few drops of dilute acid and then follow with a little dilute alkali.

A similar type of explanation will help you to account for your experimental observations.

3.3. *The reaction between chlorine and iodine*

You may have seen your teacher pass a slow stream of dry chlorine through a U-tube containing a few tiny crystals of iodine. If so, you will know that several interesting things happen. First the bottom of the tube becomes warm—a sure sign of a reaction. This is followed by the appearance of a few drops of a brown liquid and then the bend of the tube fills with brown vapour. But as the passage of the gas continued you saw small yellowish crystals form as the brown vapour disappeared.

We could summarize these observations in this way:

(a) Iodine + Chlorine → Brown liquid/vapour
(b) Brown liquid/vapour + More chlorine → Yellow crystals

If this second reaction is reversible, then the removal of some chlorine might lead to the reformation of the brown liquid.

Accordingly when we inverted the U-tube and allowed the heavy chlorine gas to fall out, we were not surprised when the yellow crystals slowly disappeared and once again we saw the brown vapour and liquid.

The cycle of changes can be repeated and so once again we are examining a reversible system.

It is not difficult to imagine that the equilibrium which we have been investigating takes place between two separate compounds formed from iodine and chlorine. It seems likely that the first compound, which you would expect to have the smaller proportion of chlorine in it, is the brown volatile liquid. Presumably the second compound, containing a higher proportion of chlorine, is the yellow crystalline solid. Analysis confirms this guess and shows that the brown liquid is iodine chloride, ICl, and the yellow solid is iodine trichloride, ICl_3.

Thus we can write the equation for this interesting pair of reactions:

$$Cl_2(g) + ICl(l) \rightleftharpoons ICl_3(s)$$

You already know that chlorine and iodine, together with bromine, are members of a chemical family called the halogens. Since two chlorine atoms readily form a chlorine molecule, Cl_2, it is not surprising that combination can also occur between a chlorine atom and a somewhat similar iodine atom, to form iodine chloride, ICl.

3.4. *Another investigation involving iodine*

If you pour a small quantity of trichloromethane (chloroform) into a test-tube and then add a few cm^3 of a solution of potassium iodide in water you will notice that the two liquids do not mix. You will see that the less dense solution of potassium iodide floats on top of the trichloromethane (Fig. 33).

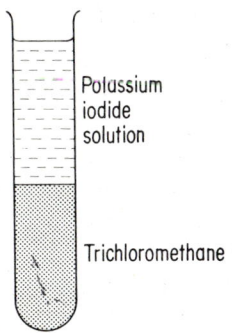

Potassium iodide solution

Trichloromethane

Fig. 33

Now take a fresh quantity of trichloromethane and add to it a small crystal of iodine. If you shake, the crystal will

dissolve and you will have the lovely purple-coloured solution.

If you next shake a similar quantity of potassium iodide solution with a second crystal of iodine the same size as the first one, you will obtain another solution, but this time it will be brown in colour.

These two solutions can be used to carry out a very interesting experiment.

With care it is possible to add some colourless potassium iodide solution to the tube containing the purple solution of iodine in trichloromethane. It is also possible to float the brown solution of iodine in potassium iodide solution on top of a colourless layer of trichloromethane.

The result of this careful manipulation is shown in the diagram (Fig. 34).

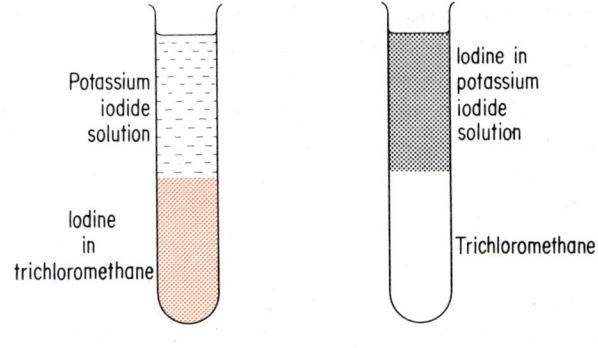

Potassium iodide solution

Iodine in trichloromethane

Iodine in potassium iodide solution

Trichloromethane

Fig. 34

If we did our work well in each tube we would have one coloured solution and one colourless (or almost colourless) liquid. The interesting thing is to observe what happens (*a*) when the two tubes are allowed to stand for some time, (*b*) when they are then gently shaken, and (*c*) when they are vigorously shaken. After this we place the two tubes side by side and try to form an impression of the degree of colour in the corresponding layers of each tube.

The results can be shown diagrammatically, see Figs. 35 and 36.

27

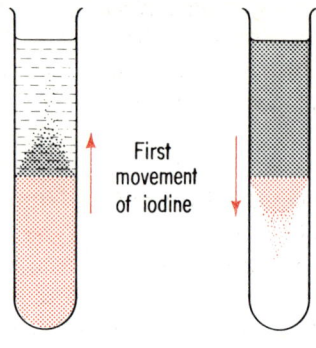

Fig. 35

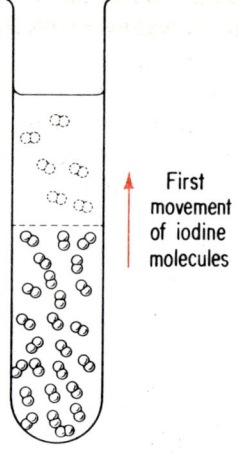

Fig. 37

Notice that as far as we can judge by external appearance the distribution of the iodine between the two layers at equilibrium is the same whichever solvent contained the iodine originally.

It is possible to carry out a further experiment to test your thinking about the equilibrium. If you take a test pipette and carefully remove the coloured upper layer of potassium iodide solution from the equilibrium mixture and replace it by fresh colourless potassium iodide solution, can you reason out what is likely to happen (Fig. 38)?

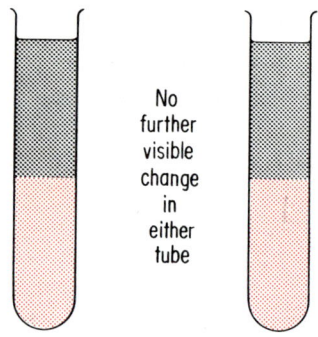

No further visible change in either tube

Fig. 36

As far as we could judge, the density of colour in the corresponding layers was the same when the experiment finished.

Can we account for what happened in these experiments?

The picture we have is of iodine particles which are dissolved in the one solute, passing, slowly at first but more rapidly with shaking, into the second liquid layer (Fig. 37). In a short while some of the iodine particles from the second layer will be passing back to the first layer. Eventually equilibrium is reached when no further external change (colour in this case) is visible, but the passage of iodine particles in both directions continues at equal and opposing rates.

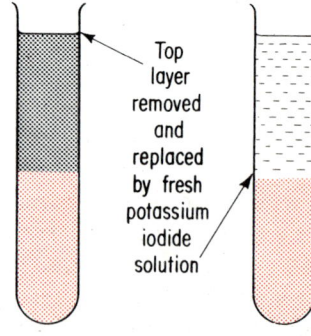

Top layer removed and replaced by fresh potassium iodide solution

Fig. 38

Think particularly what you would expect about the colour of the two layers after shaking in this case compared with the original distribution colours.

If it is possible for you to carry out this experiment, you would be able to check whether your prediction was correct.

It is worth repeating this operation three or four times, using separate fresh layers of potassium iodide solution each time.

Will it ever be possible to transfer ALL the iodine from the trichloromethane layer to the added potassium iodide solution?

3.5. *The precipitate that appears, disappears and reappears*

If you have dissolved some bismuth chloride crystals in a small quantity of concentrated hydrochloric acid you will know that, with care, you can obtain a clear colourless solution. When a few drops of this solution are poured into some water a white precipitate of bismuth chloride oxide is formed.

However, if you now add concentrated hydrochloric acid drop by drop to the white suspension the precipitate redissolves.

Here again we have a reversible reaction and the equation for the equilibrium is:

$$BiCl_3(aq) + H_2O(l) \rightleftharpoons BiOCl(s) + 2H^+(aq) + 2Cl^-(aq)$$

You will now be able to predict the effect of subsequent additions of (*a*) water, followed by (*b*) concentrated hydrochloric acid.

3.6. *The reaction between silver(I) ions and iron(II) ions*

Another reversible reaction which you may have investigated is that between silver(I) ions, $Ag^+(aq)$, and iron(II) ions, $Fe^{2+}(aq)$.

You probably first confirmed that it was possible to distinguish between pure iron(II) and iron(III) solutions by using the reagents potassium hexacyanoferrate(III), $K_3Fe(CN)_6$, and potassium thiocyanate, KCNS, and making use of the two intense colours, deep blue and deep red, shown in the table (Fig. 39).

	Potassium hexacyanoferrate(III)	*Potassium thiocyanate*
Iron(II) ions	DEEP BLUE	Colourless
Iron(III) ions	Brown	DEEP RED

Fig. 39

Why is it so unusual to have an iron(II) solution completely free from iron(III) ions?

If you then went on to mix silver(I) nitrate solution with iron(II) sulphate solution you observed a greyish precipitate which could be filtered off and identified as silver.

What test would you carry out to show that the precipitate was silver?

Thus far you would have established that:

$$Ag^+(aq) + Fe^{2+}(aq) \rightarrow Ag(s)$$

But what else is formed in the reaction?

A test of the filtrate answers the question. The solution contains iron(III) ions.

Hence the completed reaction is:

$$Ag^+(aq) + Fe^{2+}(aq) \rightarrow Ag^+(s) + Fe^{3+}(aq)$$

We can easily find out if this reaction is reversible by taking some fresh iron(III) solution, which we show to be free from iron(II) ions, and shaking it with some of the carefully washed silver which we obtained in the previous experiment.

What do you expect to happen to the silver—or at least to some of it? What colour do you expect to see when a solution of potassium hexacyanoferrate(III) is added to some of the supernatant liquid?

In this way you have established the reversible nature of the reaction so that we can write:

$$Ag^+(aq) + Fe^{2+}(aq) \rightleftharpoons Ag(s) + Fe^{3+}(aq)$$

Up to now, as a result of our investigations, we have shown that when we are dealing with a reversible change the reaction can be made to go first one way and then the other. In the case of this reaction we can find out whether both the products and the reactants are present at the same time. This would mean carrying out tests on an equilibrium mixture to see if there is evidence for the presence of both iron(II) ions and iron(III) ions in the solution.

If you were able to do this you found out that both of these ions were present and thus you established, what we have pictured all along, that at equilibrium we have a system in which the reactants and the products are all present together.

3.7. *Testing our ideas of dynamic equilibrium*

Though we have been thinking chiefly of dynamic equilibrium in a chemical system, the idea is equally valuable in giving a helpful picture of what happens in a saturated solution.

We know that sodium chloride crystals will dissolve in water, and we also know that, under certain conditions, a saturated solution of sodium chloride will deposit crystals. Here is another two-way movement and hence we have the possibility of an equilibrium situation.

Consider a sodium chloride crystal standing in a saturated solution of sodium chloride (Fig. 40). To simplify the diagram no attempt has been made to represent the groups of water molecules which cluster around the ions in solution.

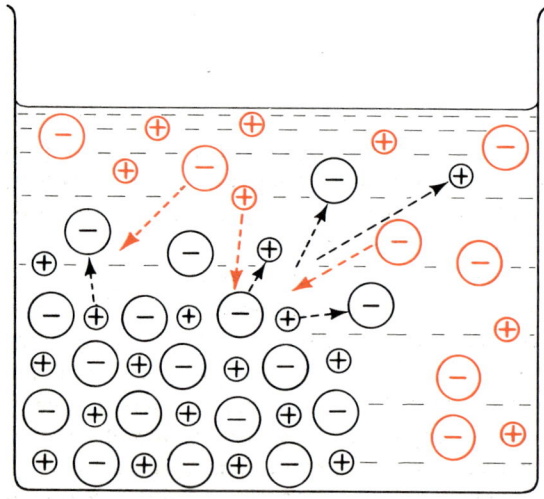

Fig. 40

It is possible that some sodium ions and some chloride ions from the surface of the crystal might move off into the solution and thus momentarily upset the balance existing at saturation point. As a result an equal number of sodium ions and chloride ions would attach themselves to the crystal surface so that the equilibrium was restored.

So far this is only guesswork. To check our guess we need a way of labelling some of the ions which we know to be originally in the undissolved solid. It is now possible to do this by using radioactive materials.

Lead(II) chloride can be prepared which contains some radioactive lead ions. We can then add some of this to an already saturated solution of non-radioactive lead chloride. Since the solution is already saturated, then as long as we keep the temperature constant, IN TOTAL no more lead chloride can dissolve. (This is what we mean by saying a solution is saturated at a given temperature.)

If we shake the radioactive solid with the non-radioactive saturated solution and then filter and test the filtrate, we find that it contains radioactive lead ions in solution. This means that radioactive lead ions have passed from the solid to exchange with some non-radioactive lead ions which were previously in solution. It follows that this traffic must be a two-way movement and thus some non-radioactive lead ions must have deposited on the solid.

This is strong supporting evidence for our collision hypothesis and our picture of dynamic equilibrium at the particle, in this case ionic, level.

3.8. *A final example of chemical equilibrium*

At some point in your practical work when you were carrying out experiments with nitric acid or nitrates, you probably met a brown gas, sometimes called nitrogen dioxide—or less accurately, nitrogen peroxide.

If two identical glass vessels are filled with samples of this gas from the same generator at the same pressure, they can be used for a dramatic demonstration of chemical equilibrium.

One of the two sealed bulbs can be kept as the control for colour comparison purposes. At room temperature, both bulbs

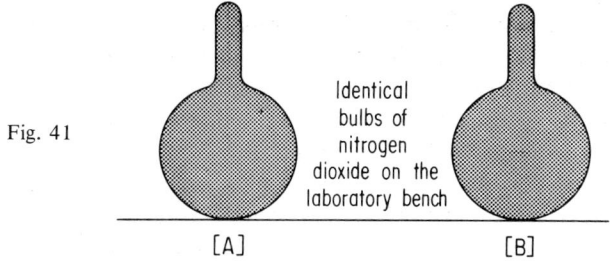

Fig. 41

Identical bulbs of nitrogen dioxide on the laboratory bench

[A]   [B]

as they stand side by side on the laboratory bench will appear a reddish brown colour (Fig. 41). If one bulb (A) is now placed in a beaker of ice water the colour will be seen to change, becoming gradually pale yellow and finally almost colourless (Fig. 42).

If bulb A is now removed from the ice water and allowed to warm up again to room temperature, its colour darkens and returns to the original shade of brown, identical with that in the control bulb B. To ensure that there is nothing special about bulb A you could repeat the experiment, this time cooling bulb B and keeping bulb A as your control colour. You will find the result is the same.

Having established the repeatability of the cooling change, now place one of the bulbs in a beaker of boiling water and allow time for the temperature inside the bulb to become

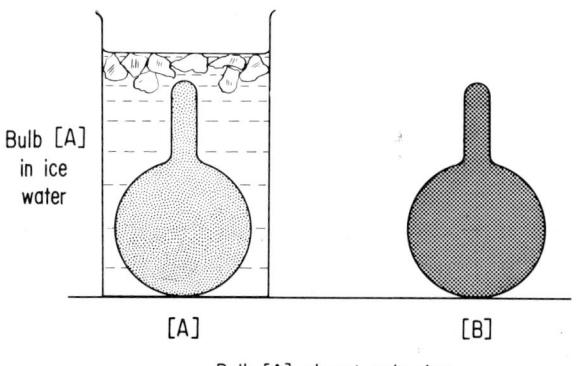

Bulb [A] in ice water

[A]   [B]

Bulb [A] almost colourless

Fig. 42

steady. Again we see a marked change in colour; this time the hot bulb darkens appreciably until it appears a very dark brown (Fig. 43). Once again, if we allow the hot bulb to cool back to room temperature, the original colour returns. In the same way as before, it can easily be shown that either bulb will undergo this reversible cycle of colour changes. By controlling the heating we can change the colour of the gas in the

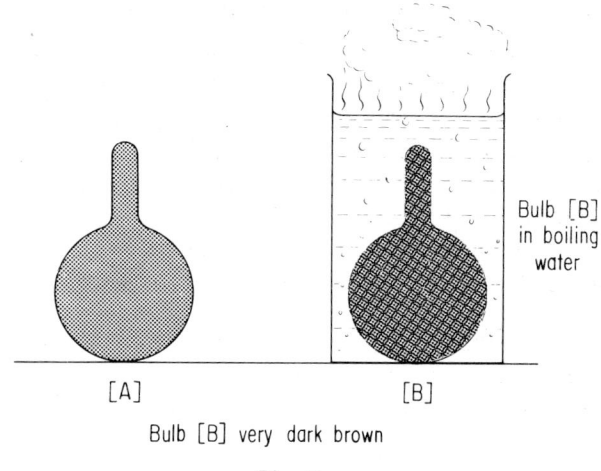

Bulb [B] in boiling water

[A]   [B]

Bulb [B] very dark brown

Fig. 43

bulb from almost colourless to almost black, stopping at any intermediate colour by stopping at an intermediate temperature.

This is an experimental result, but chemists will not be satisfied until they have some reasonable explanation to account for the observed changes.

The explanation proves to be quite simple.

By a series of measurements, which you will learn about later in your study of chemistry, it is possible to show that the pale yellow gas at 0 °C consists almost entirely of dinitrogen tetraoxide molecules, N_2O_4, whereas the dark brown gas at 100 °C consists largely of nitrogen dioxide molecules, NO_2.

How do you account for the intermediate colours?

We are clearly dealing with a reversible reaction which only involves these two molecular species and the equilibrium is given by the equation:

$$N_2O_4(g) \rightleftharpoons 2NO_2(g)$$

The explanation of the deepening colour on heating is the steadily increasing dissociation of dinitrogen tetraoxide molecules, which are colourless, resulting in the formation of nitrogen dioxide molecules, which are dark brown in colour. At any intermediate temperature, differing mixtures of dark brown and colourless molecules can obviously produce a whole range of intermediate colours.

When a bulb of gas is cooled, the reverse change occurs in which molecular combination leads to the increasing formation of dinitrogen tetraoxide and accordingly the colour lightens.

Two such bulbs of gas can be used to demonstrate conclusively that the particular equilibrium mixture which we obtain is exactly the same whether we approach it from a higher temperature or from a lower temperature. The composition of the equilibrium mixture, that is, what proportion of it is made up of each type of molecule, depends only on the temperature, provided the pressure of the two gases is the same. In more advanced work you will consider what happens if we keep the temperature constant in a bulb, but alter the pressure of the gas inside. Some of you may want to point out that during the heating of the bulbs there must have been some pressure increase. This is true, but if the temperature rise is not too great the pressure increase will be relatively small.

3.9. Chemical equilibria and industry

We might end our brief consideration of chemical equilibria by recalling our box model with the light balls being blown from side to side.

If we have a high barrier and a low blower pressure, then very few balls will pass over. The requirements for a ball to cross the barrier from one compartment to the other are:

(a) That it have sufficient energy to reach the required height.

(b) That it be moving in tne right direction.

32

When we increase the blower pressure a larger proportion of balls possess these two necessary requirements, and hence more balls pass over the barrier. We have compared these requirements to the activation energy of a chemical reaction, and as in our model we could easily alter the barrier height, so in different chemical reactions the minimum necessary energy to initiate them varies considerably.

We have discussed earlier the equilibrium situation, first with equal air pressure on both sides of the barrier and second with one pressure much greater than the other. The latter situation is important, for it reminds us that in most reversible chemical reactions the activation energy for the reverse reaction is not the same as that for the forward reaction.

It would be possible to construct a modified box (Fig. 44) in which the base level of compartment B is lower than that of compartment A. We would thus be simulating a situation in

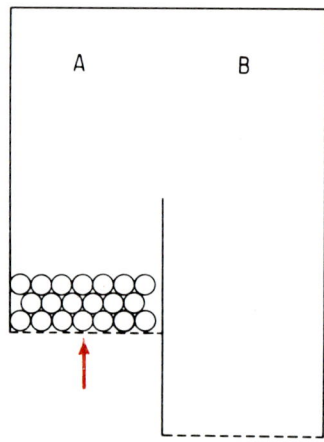

Fig. 44

which the energy barrier to be surmounted by balls returning from B to A would be greater than that confronting balls attempting to pass from A to B. Yet if we did this, some balls would still pass from B to A and we would not end up with all the balls in compartment B, even though we might well have far more on that side when the equilibrium was reached. This situation would represent an equilibrium position with the balance point very much nearer one side than the other, a condition which we always strive to achieve when we use a reversible reaction to prepare a desired product.

If a chemical is to be manufactured by the use of a reversible reaction, it is vital that the process be as economic as possible. This means:

(*a*) We need to have the greatest convenient conversion of the starting materials into the final product.

(*b*) This maximum conversion must take place as rapidly as possible.

Both of these requirements involve the application of principles which you have studied in this book, the importance of which should now be abundantly clear.

Basic Questions

B1. You have probably carried out an experiment in which you have heated magnesium in a crucible. Which of the following would you expect to influence the rate at which the magnesium reacts with the air?
(a) The temperature of the air.
(b) The temperature of the magnesium.
(c) The shape of the crucible.
(d) The air pressure.
(e) The concentration of oxygen in the air.
(f) The surface area of the magnesium.

B2. (a) If you have answered question B1, you will have chosen certain factors which you feel might influence the rate at which magnesium reacts with air. Explain briefly why these factors do influence the rate of reaction.
(b) Can you suggest any other factors which might influence the rate of this reaction?

B3. Which of the following factors would you expect to influence the initial rate at which carbon dioxide is produced when calcium carbonate reacts with an acid? The acid is always present in sufficient quantity to cover the calcium carbonate.
(a) The mass of calcium carbonate.
(b) The surface area of the calcium carbonate.
(c) The material of the reaction vessel.
(d) The concentration of the acid.
(e) The size of the reaction vessel.
(f) The state of division of the calcium carbonate.
(g) The quantity of acid.
(h) The temperature of the acid.
(i) The solubility of the reaction products.

B4. (a) Consider the nine factors given in Question B3 and decide which ones would influence the total mass of carbon dioxide produced in the reaction.
(b) Can you suggest any other factors which might influence the initial rate of this reaction?

B5. It is known that the thermal decomposition of potassium chlorate into potassium chloride and oxygen is catalysed by certain metal oxides. Describe concisely experiments you would carry out to decide whether iron(III) oxide catalyses this reaction. Your experiments should indicate how you would decide.
(a) Whether the reaction is more rapid at a constant temperature when iron(III) oxide is present.
(b) Whether the iron(III) oxide is consumed in the reaction.
(c) Whether the reaction products are changed by the addition of iron(III) oxide.
(Of the substances involved in the reaction only iron(III) oxide is insoluble in water.)

B6. It is known that under certain conditions hydrogen peroxide decomposes rapidly into water and oxygen. Devise a method to determine how rapidly the oxygen is evolved in this reaction.
How would you use your method to determine:
(a) Whether manganese(IV) oxide catalyses this reaction?
(b) Whether the rate of the reaction depends on the quantity of manganese(IV) oxide?
(c) Whether the rate of the reaction depends on the state of division of the manganese(IV) oxide?
(d) Whether, with manganese(IV) oxide present, the rate of reaction is influenced by temperature?
(e) Whether zinc(II) oxide is a better catalyst than manganese(IV) oxide?

B7. You have probably seen the reaction of dilute sulphuric acid on zinc which produces hydrogen gas and a solution of zinc(II) sulphate. Devise a method to determine how rapidly the hydrogen is evolved in the reaction.
How would you use your method to determine to what extent the initial rate of this reaction is influenced by (a) the amount of zinc, (b) the state of division of the zinc, (c) the temperature of the reaction mixture, (d) the presence of copper sulphate, and (e) the concentration of the sulphuric acid.

34

B8. Attempt to give simple explanations for the following observations.

(a) Wire-wool will burn readily if placed for a moment in a bunsen flame, but it is not possible to make a nail burn by the same method.

(b) It is difficult to ignite a log of wood, but a wooden splint will ignite easily.

(c) On a bottle of fine aluminium powder is found the warning 'Highly inflammable, dust explosion possible'.

(d) A mixture of hydrogen and oxygen is stable at room temperature, but if a cold specimen of a certain metal is put in the gas mixture an explosion can be produced.

(e) A mixture of hydrogen and chlorine is stable in the dark, but in sunlight a reaction takes place.

B9. A small copper sphere was put in a boiling tube, covered with dilute nitric acid at 30 °C and gently shaken. There was a slow reaction to produce a blue solution and bubbles of gas.

Arrange the following in probable order of decreasing effectiveness in influencing the rate of the reaction (most effective first):

(a) Doubling the concentration of the nitric acid.

(b) Doubling the quantity of nitric acid.

(c) Doubling the temperature to 60 °C.

(d) Doubling the volume of the copper sphere.

(e) Doubling the rate of shaking the reaction mixture.

B10. A pupil is investigating the reaction between sodium(I) hydroxide solution and potassium permanganate solution (purple). Among the products of the reaction are green potassium manganate solution and oxygen gas.

The pupil wants to discover how the rate of the reaction is influenced by the concentrations of the reactants. To do this he needs to know how far the reaction has progressed at various times after mixing the reactants.

All the following would be possible ways of determining the extent of the reaction, but ONE would not be used by the pupil to solve his particular problem. Which one is it?

(a) The quantity of oxygen produced.

(b) The decrease in the purple colour.

(c) The increase in the green colour.

(d) The temperature change.

(e) The decrease in mass.

(f) The change in electrical conductivity.

(g) The change in the alkalinity of the solution.

B11. A pupil is investigating the decomposition of a substance in solution. Here is some information about the reaction:

One of the products is a gas.
One of the products is an insoluble solid.
The reaction is very exothermic.
Decomposition at room temperature is not rapid.

(A) Suggest methods which the pupil might use to determine the rate of the reaction.

(B) Which ONE of the following will be the most probable effect of warming the reaction mixture with a small bunsen flame of constant size?

(a) Decomposition will stop.

(b) Decomposition will be unchanged in rate because one of the products is escaping.

(c) Decomposition will speed up, but less rapidly than one might expect because the reaction is exothermic.

(d) Decomposition will speed up, and more rapidly than one might expect because the reaction is exothermic.

(e) Decomposition will slow down due to greater heat loss at higher temperatures.

B12. One of the following, (a)–(h), is a better description of a catalyst than any of the others. Which ONE is the most satisfactory description?

(a) A substance which increases the rate of reactions.

(b) A substance which increases the rate of some reactions.

(c) A substance which can start reactions that would not take place without the catalyst.

(d) A substance which can alter the rate of some reactions.

(e) A substance which can alter the rate of some reactions without being changed in any way.

(f) A substance which can alter the rate of some reactions without itself being changed, except perhaps in mass.

35

(g) A substance which can increase the rate of some reactions without being changed in mass.

(h) A substance which can decrease the rate of some reactions without being changed in any way.

B13. Equal masses (0.4 g) of fresh calcium turnings were reacted in separate experiments with excess water and dilute
hydrochloric acid. The volume of hydrogen liberated was
measured at half-minute intervals with a syringe, and these
volumes were corrected to s.t.p. The results obtained are
given in the table:

Time (minutes)	Volume of hydrogen produced using water (cm³)	Volume of hydrogen produced using dilute HCl (cm³)
0.5	17	88
1.0	34	144
1.5	76	182
2.0	134	207
2.5	184	222
3.0	216	224
3.5	220	224
4.0	222	224

(a) Plot the results in a suitable manner on graph paper.
(b) Which reaction has the greater initial rate?
(c) In which reaction is the rate greater at the two-
minute time?
(d) Is there any time at which the rates of the two reactions
are the same?
(e) What mass of calcium will be left after one minute in
each case?
(f) Calculate the mass of calcium that will liberate 1 mole
of hydrogen (22.4 litres at s.t.p.).
(g) Find the atomic mass of calcium and compare this
with the answer to (f). Can you account for the relation between the values?
(h) Suggest an explanation for the answer to (b).

The following ten questions are concerned with an
investigation made by a pupil of the reaction between molar
hydrogen peroxide and manganese(IV) oxide. He is interested
in the way the rate of the reaction at the start depends on
various factors, and he varies only one of these at a time.

Some possible ways in which the initial rate and the final
volume of oxygen may be influenced are shown below.

	Effect on rate	Effect on final volume of oxygen
(a)	unchanged	unchanged
(b)	unchanged	increase
(c)	unchanged	decrease
(d)	increase	increase
(e)	increase	unchanged
(f)	increase	decrease
(g)	decrease	decrease
(h)	decrease	increase
(i)	decrease	unchanged

Which one of the effects would you expect in each case if
the pupil:

B14. Added to the original hydrogen peroxide an equal volume
of 5 molar hydrogen peroxide?

B15. Added to the original hydrogen peroxide an equal volume of
0.1 molar hydrogen peroxide?

B16. Raised the temperature?

B17. Lowered the temperature?

B18. Used the same quantity of manganese(IV) oxide but in a
finer state of division?

B19. Added to the original hydrogen peroxide an equal volume of
water?

B20. Used 0.1 molar hydrogen peroxide instead of molar?

B21. Used 2 molar hydrogen peroxide instead of molar?

B22. Carried out the original reaction in a larger vessel?

B23. Carried out the original reaction in a centrifuge at 2000
revs/min?

B24. Which of the following changes would you regard as
reversible, and which as irreversible?
(a) Heating zinc(II) oxide.
(b) Burning magnesium in air.

(c) Heating mercury in air.
(d) Heating red lead oxide in air.
(e) Converting liquid water to steam.
(f) Heating mercury(II) oxide.
(g) Melting ice.
(h) Heating calcium carbonate.
(i) Driving off water from copper(II) sulphate crystals.
(j) Passing steam over strongly heated iron.

B25. Explain why it is possible for iron, steam, iron(II) diiron(III) oxide and hydrogen to exist in dynamic equilibrium under certain conditions, while when steam is passed over iron in a strongly heated tube there is complete conversion of the iron to iron oxide.

B26. You will probably have read about the experiment in which a mixture of iron, steam, iron(II) diiron(III) oxide and hydrogen is heated in a closed vessel. After some time it is found that the amounts of the four substances do not change any further, and the system is said to be in 'dynamic equilibrium'. Which one of the following is the best description of the state of dynamic equilibrium?
(a) All reaction has stopped.
(b) Reaction is taking place in both directions.
(c) Reaction is taking place at equal rates in both directions.
(d) Reaction is taking place very slowly in both directions.
(e) Reaction is taking place very rapidly in both directions.

B27. When hydrogen sulphide solution reacts with zinc ions in solution to precipitate zinc(II) sulphide, the following equilibrium is established.

$$Zn^{2+}(aq) + H_2S \rightleftharpoons ZnS(s) + 2H^+(aq)$$

Which of the following would probably favour the formation of zinc(II) sulphide?
(a) Addition of extra zinc ions.
(b) An increase in the concentration of dissolved hydrogen sulphide.
(c) An increase in the acidity of the solution.
(d) Addition of sodium hydroxide.
(e) Addition of hydrogen chloride gas to the solution.

B28. You probably met bromine, a dark red liquid, when you discussed the halogen family of elements (Nuffield topic 13.3). When excess bromine is shaken with water in a corked container, the water becomes orange in colour. At first the orange colour deepens on continued shaking, but after about five minutes the intensity of the orange colour remains constant even though there is liquid bromine remaining. Which one of the following explanations, (a)–(g), best fits the situation in the flask after five minutes (for 'halogen' read 'bromine')?
(a) Some halogen molecules are of a type which can dissolve, while others are not. All the molecules of the dissolving type have dissolved in the first five minutes.
(b) The halogen molecules go into solution rapidly at first and then more slowly, until after five minutes the rate of going into solution is so slow that there is no further change in colour.
(c) The halogen molecules are going into solution rapidly, and after five minutes are also coming out of solution rapidly. This restricts any change in colour to the first five minutes.
(d) The halogen molecules are coming out of solution so slowly after five minutes that no further change in colour is observed.
(e) After five minutes the halogen molecules are going into solution and leaving the solution at the same rate.
(f) After five minutes the halogen molecules are going into solution and leaving the solution so slowly that no change in colour is observed.
(g) As the concentration of the halogen molecules in the water increases, more halogen molecules can escape as gas into the space above the solution. After five minutes the halogen molecules are escaping into the gas as rapidly as halogen is going into solution.

B29. When excess iodine, a dark solid, is shaken with warm water in a corked flask the water becomes pale orange, and the colour deepens until after about five minutes there is no further change. Which one of the explanations of question B28 best fits the situation in the flask after five minutes (for 'halogen' read 'iodine')?

37

B30. *Excess* iodine is shaken with 10 cm³ water and 5 cm³ of benzene (in which iodine is appreciably soluble) in a corked boiling tube until no more iodine will dissolve in either solvent. On standing, the water and benzene form two layers, the benzene floating on the water. The colour of each layer is noted, this colour being a measure of the iodine concentration in the layer. Note that excess iodine is used. There is solid iodine left at the bottom of the boiling tube.

5 cm³ more benzene is added to the mixture, and the tube is shaken until equilibrium has been reached again. Which ONE of the following sets of changes is most likely to occur:

	Change in amount of solid iodine	Change in equilibrium colour of water layer	Change in equilibrium colour of benzene layer
(*a*)	decrease	increase	increase
(*b*)	decrease	no change	no change
(*c*)	no change	no change	no change
(*d*)	no change	increase	decrease
(*e*)	decrease	increase	no change

B31. *Excess* iodine is shaken with 10 cm³ water and 5 cm³ benzene in a corked boiling tube until no more iodine will dissolve in either solvent. The equilibrium colour of each layer is noted.

5 cm³ more water is added to the mixture and the tube is shaken until equilibrium has been reached again. Which ONE of the sets of changes listed in Question B30 is most likely to occur to the amount of solid iodine, the colour of the water layer and the colour of the benzene layer?

B32. *Excess* solid iodine is shaken with 10 cm³ water and 5 cm³ benzene in a corked boiling tube until no more iodine will dissolve in either solvent. The equilibrium colour of each layer is noted.

5 cm³ of water is then removed from the tube and the remaining mixture is shaken until equilibrium has been reached again. Which one of the sets of changes listed in Question B30 is most likely to occur to the amount of solid

iodine, the colour of the water layer and the colour of the benzene layer?

B33. It is known that iodine is more soluble in potassium bromide solution that it is in pure water. Potassium bromide dissolves in water but not in benzene.

Excess solid iodine is shaken with 10 cm³ of water and an equal volume of benzene in a corked boiling tube until no more iodine will dissolve in either solvent. The equilibrium colour of each layer is noted.

A few crystals of potassium bromide are added and the mixture is shaken until equilibrium has been reached again. Which one of the sets of changes listed in Question B30 is most likely to occur to the amount of solid iodine, the colour of the water layer and the colour of the benzene layer?

B34. Excess solid iodine is shaken with 10 cm³ of water and an equal volume of benzene until no more iodine will dissolve in either layer. The equilibrium colour of each layer is noted.

The temperature of the mixture is then increased by about 20 K and the shaking is continued until equilibrium has been reached again. Which one of the sets of changes listed in Question B30 is most likely to occur to the amount of solid iodine, the colour of the water layer and the colour of the benzene layer?

B35. Antimony(III) chloride is a white solid which dissolves in concentrated hydrochloric acid to give a clear colourless solution. When this solution is poured into water a white precipitate of antimony(III) chloride oxide appears.

The white precipitate dissolves if sufficient concentrated hydrochloric acid is added. The reaction is a reversible one and can be represented by the equation

$$SbCl_3(aq) + H_2O(l) \rightleftharpoons SbOCl(s) + 2H^+(aq) + 2Cl^-(aq)$$

(*a*) Consider the solution of $SbCl_3$ in concentrated hydrochloric acid. Predict the probable effect of adding to it in separate experiments (i) water, (ii) concentrated hydrochloric acid, (iii) concentrated nitric acid.

(*b*) Consider the white precipitate of SbOCl produced by

adding water to a solution of SbCl$_3$ in concentrated hydrochloric acid. Predict the probable effect of adding to this white precipitate in separate experiments (i) water, (ii) concentrated hydrochloric acid, (iii) concentrated nitric acid.

Revision Questions

Questions R1 to R11 are concerned with a series of experiments carried out by a pupil to investigate the reaction between hydrogen peroxide and cobalt(II) sulphate.

R1. Which of the following statements about cobalt sulphate will probably be correct?
   (*a*) It is coloured.
   (*b*) There is more than one cobalt sulphate.
   (*c*) It will catalyse some reactions.
   (*d*) It is a compound of a transition element and two non-transition elements.
   (*e*) The metal from which it is formed is less reactive than the alkali metals.
   (*f*) It will contain water of crystallization.

R2. You may have studied the decomposition of hydrogen peroxide in the presence of manganese(IV) oxide. The pupil was interested to know whether cobalt sulphate would also catalyse the decomposition.

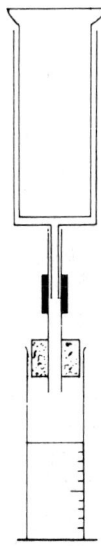

The pupil took 200 cm^3 of neutral '10 volumes hydrogen peroxide' in a flask and to it added 10 cm^3 of cobalt sulphate solution. He estimated the hydrogen peroxide concentration at various intervals by withdrawing a sample in a measuring cylinder, adding a quantity of manganese(IV) oxide to give rapid decomposition of any remaining hydrogen peroxide, and rapidly attaching a syringe as shown in the sketch to determine the volume of oxygen evolved.

If '10 volumes hydrogen peroxide' means that 1 volume of hydrogen peroxide can be decomposed to 10 volumes of oxygen under room conditions, which of the following volumes of oxygen would the pupil have obtained if he had taken a 10 cm^3 sample of the mixture immediately after adding the cobalt sulphate solution?

(a) 10 cm^3, (b) 95 cm^3, (c) 100 cm^3, (d) 105 cm^3, (e) 210 cm^3, (f) 2000 cm^3.

R3. Here is a graph to show the results the pupil obtained when he took several samples of volume x cm^3 each time at various intervals after adding the cobalt sulphate.

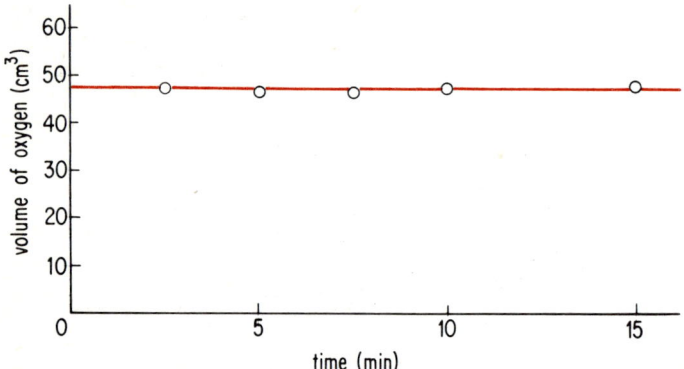

If x had one of the following values, which was it?

(a) 1, (b) 2, (c) 5, (d) 10, (e) 47, (f) 50, (g) 200, (h) 210.

R4. Only one of the following observations about the graph in Question R3 is correct. Which one is correct?

(a) The rate of the reaction is directly proportional to the hydrogen peroxide concentration.

(b) The rate of the reaction is directly proportional to the catalyst concentration.

(c) The rate of the reaction does not depend on the hydrogen peroxide concentration.

(d) The rate of the reaction does not depend on the catalyst concentration.

(e) Cobalt sulphate is not a catalyst under these conditions.

R5. Here is a set of results the pupil obtained by taking separately 200 cm^3 of 10 volumes hydrogen peroxide, 10 cm^3 of cobalt sulphate solution and 1 cm^3 of sodium hydroxide solution.

The pupil mixed the cobalt sulphate and the sodium hydroxide. This produced a dark precipitate of an insoluble cobalt compound. The pupil waited a short time and then tipped the mixture into the hydrogen peroxide.

The pupil took 10 cm^3 portions of the reaction mixture at various times and determined the amount of hydrogen peroxide remaining by the method of oxygen evolution described in Question R2.

Times after mixing (minutes)	Volume of oxygen (cm^3)
1	91
2	87
3	83
4	79
5	75

Draw a graph of volume of oxygen (y axis) against time and predict the volume of oxygen that would have been produced by 10 cm^3 of reaction mixture after ten minutes.

R6. Only one of the following statements follows directly from the graph drawn in Question R5.

(a) The rate of the reaction is directly proportional to the hydrogen peroxide concentration.

(b) The rate of the reaction is directly proportional to the catalyst concentration.

(c) The rate of the reaction is independent of the hydrogen peroxide concentration.

(d) The rate of the reaction is independent of the catalyst concentration.

(e) The straight-line graph indicates that the reaction is not being catalysed.

(*f*) The rate of the reaction is directly proportional to the amount of sodium hydroxide used.

(*g*) The rate of the reaction is directly proportional to the volume of solution remaining.

R7. The pupils repeated the experiment, using the same volumes as in Question R5, but this time he first mixed the hydrogen peroxide and cobalt sulphate, and then added the sodium hydroxide. He took 10 cm³ portions of the mixture at various times after mixing as in the previous experiment.

Time after mixing (minutes)	Volume of oxygen (cm³)
0.5	66
1	57
2	48
3	40
4	32
5	24

Plot a graph of volume of oxygen (*y* axis) against time. What would the volume of oxygen have been if the pupil had taken a 10 cm³ sample at the moment after mixing?

R8. Only one of the following statements follows directly from the graph drawn in Question R7.

(*a*) The rate of the reaction is directly proportional to the hydrogen peroxide concentration.

(*b*) The rate of the reaction is directly proportional to the catalyst concentration.

(*c*) The rate of the reaction is independent of the hydrogen peroxide concentration.

(*d*) The rate of the reaction is independent of the catalyst concentration.

(*e*) During the first minute the rate of the reaction is directly proportional to the hydrogen peroxide concentration.

(*f*) During the first minute the rate of the reaction is directly proportional to the catalyst concentration.

(*g*) During the first minute the rate of the reaction is independent of the hydrogen peroxide concentration.

(*h*) During the first minute the rate of the reaction is independent of the catalyst concentration.

(*i*) Between the first and the fifth minute the rate of the reaction is directly proportional to the hydrogen peroxide concentration.

(*j*) Between the first and the fifth minute the rate of the reaction is directly proportional to the catalyst concentration.

(*k*) Between the first and the fifth minute the rate of the reaction is independent of the hydrogen peroxide concentration.

(*l*) Between the first and the fifth minute the rate of the reaction is independent of the catalyst concentration.

R9. Look at the curves you have obtained for Questions R5 and R7. In which case is the rate of reaction greater: (*a*) at the beginning, (*b*) at the two-minute stage?

R10. Can you put forward any explanation for (*a*) the change in rate during the first minute for the reactions of Questions R5 and R7, (*b*) the difference between the rates of the same reactions at the two-minute stage?

R11. Here are some curves obtained at 20 °C with (*a*) hydrogen peroxide and cobalt(II) sulphate solution, (*b*) hydrogen

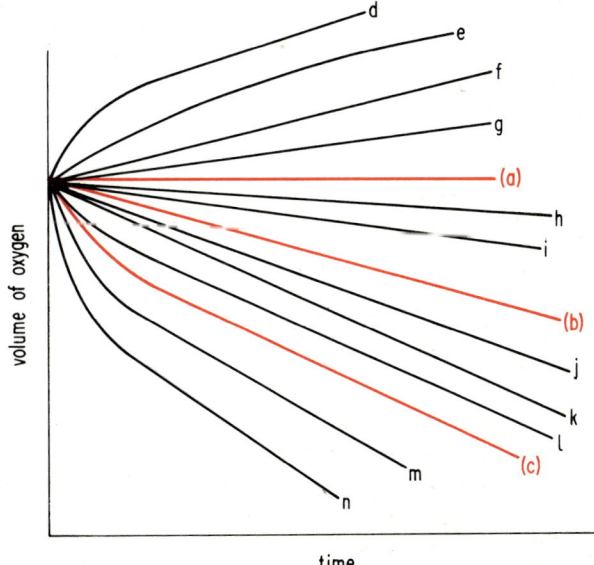

41

peroxide to which was added a mixed solution of cobalt sulphate and sodium(I) hydroxide, (c) a mixed solution of hydrogen peroxide and cobalt sulphate to which was added sodium(I) hydroxide solution.

You will also find curves lettered d to n. In each of the following cases, which ONE of the curves would be most likely to represent (i) (a) at 30 °C, (ii) (a) at 40 °C, (iii) (b) at 30 °C, (iv) (b) at 40 °C, (v) (c) at 30 °C, (vi) (c) at 40 °C?

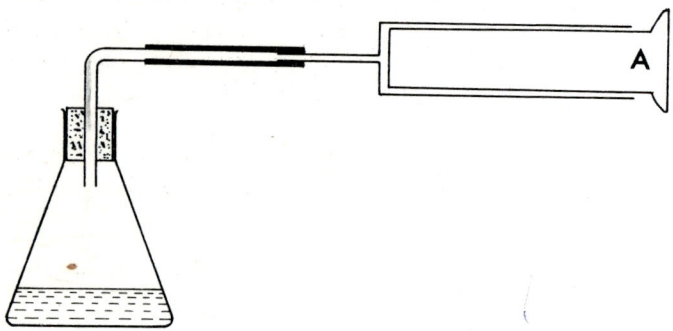

Questions R12–R18 are concerned with the investigation made by a pupil into the reaction between powdered lead(II) carbonate and a large excess of dilute nitric acid. The reaction was carried out in the flask shown in the diagram, and the volume of carbon dioxide produced was determined by the large horizontal syringe.

The pupil took 0.01 mole of lead(II) carbonate and found that eventually 230 cm³ of carbon dioxide was produced (this volume being corrected to s.t.p. where 1 mole would occupy 22.4 litres).

R12. A chemical reaction involves small whole numbers of molecules. The pupil could reasonably suppose that the volume of carbon dioxide should have been 224 cm³ (as 230 is so close to 224). If there is only one reason for this discrepancy it is most likely to be because:
(a) The syringe was sticking.
(b) Carbon dioxide dissolves in the reaction mixture.

(c) The lead(II) carbonate was damp.
(d) The pupil took too much lead(II) carbonate.
(e) There was a slight leak in the apparatus.
(f) The syringe was incorrectly graduated.

R13. From the figures given, the pupil could begin to work out the equation for the reaction. This seems to be close to one of the following:
(a) $2PbCO_3 + $ nitric acid $\rightarrow CO_2 +$
(b) $PbCO_3 + $ nitric acid $\rightarrow 2CO_2 +$
(c) $PbCO_3 + $ nitric acid $\rightarrow CO_2 +$
(d) $0.01PbCO_3 + $ nitric acid $\rightarrow 230CO_2 +$
(e) $0.01PbCO_3 + $ nitric acid $\rightarrow 224CO_2 +$

R14. If the pupil raised the syringe into the vertical position (with end A uppermost) during the reaction, which one of the following is most likely to have happened?
(a) The reaction stopped almost at once.
(b) The cork blew out of the flask.
(c) The final volume was slightly greater than before.
(d) The reaction took place more rapidly.
(e) The final volume was slightly less than before.

R15. If the pupil lowered the syringe into the vertical position (with end A lowest) during the reaction, which one of the following is most likely to have happened?
(a) The barrel dropped out of the syringe.
(b) The final volume was slightly greater than before.
(c) The reaction mixture was sucked into the syringe.
(d) The final volume was slightly less than before.
(e) The reaction stopped.
(f) The reaction never reached completion.

R16. The pupil used 0.005 mole of lead(II) carbonate while keeping the quantity of nitric acid unchanged. The number of cm³ of carbon dioxide obtained (corrected to s.t.p.) was close to one of the following (a) 23, (b) 46, (c) 115, (d) 224, (e) 230, (f) 460, (g) 1150.

R17. The pupil then used 0.01 mole of lead(II) carbonate and twice as much nitric acid as in the original experiment.

The number of cm³ of carbon dioxide he obtained was close to one of the following (a) 115, (b) 230, (c) 460, (d) 920, (e) 1150.

R18. If in the original experiment the pupil had used 0.01 gramme formula of lead(II) carbonate in lump form and excess dilute nitric acid, he would have obtained one of the following results:

(a) The reaction would have been more rapid at the start and the final volume would have been greater.

(b) The reaction would have been slower at the start and the final volume would have been smaller.

(c) No change in either the rate at the start or in the final volume.

(d) The reaction would have been slower at the start and the final volume would have been unchanged.

(e) The reaction would have been slower at the start and the final volume would have been greater.

R19. The next five questions are concerned with the following equilibria which can be rapidly established in aqueous solution:

$$\text{Iodine solid} \xrightleftharpoons[]{\text{water}} \underset{\text{(brown)}}{\text{Iodine solution}} \xrightleftharpoons[H^+]{'OH^-} \text{Colourless solution of iodine salts}$$

Consider the case where there is a large excess of solid iodine in contact with neutral water. Which ONE of the following changes will take place in the equilibrium colour of the solution when a few drops of sodium hydroxide solution is added?

(a) The brown solution will become deeper brown.

(b) The brown solution will become less brown.

(c) The brown solution will become colourless.

(d) No change in the brown colour of the solution.

(e) Probably some change in colour but it is not possible to decide which way.

(f) The colourless solution will become brown.

(g) The colourless solution will remain colourless.

R20. Which ONE of the changes listed in Question R19 will occur to the equilibrium mixture containing a few drops of sodium hydroxide solution when more than enough acid is added to neutralize the alkali?

R21. Now consider the case where there is only a small quantity of solid iodine in contact with neutral water. Which ONE of the changes listed in Question R19 will take place to the equilibrium mixture on adding a large excess of sodium hydroxide solution?

R22. When one drop of sodium hydroxide solution is added to an iodine solution (no solid iodine present) the brown colour decreases. What further change will occur to the colour of this solution if its temperature is then raised? Choose ONE of the changes listed in Question R19.

R23. This question is more difficult than numbers R19–R22. Attempt it only if you have been able to answer the previous four questions.

In answering Questions R19–R22 you will have used the information about the equilibria given at the beginning of Question R19.

The 'colourless solution of iodine salts' is a better solvent than water for solid iodine. The equilibria present in an alkaline solution could be represented in the way shown below:

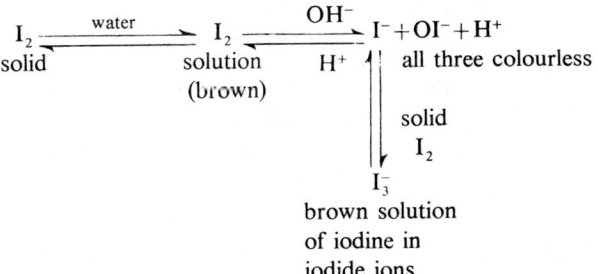

Look at Questions R19–R22 in the light of this new information and decide whether any changes will be required in your answers to these four questions. You are not meant to use this additional information in obtaining your original answers to Questions R19–R22.

More Searching Questions

MS1. You have probably carried out an experiment to find how the rate of production of sulphur in the reaction between sodium thiosulphate and dilute hydrochloric acid depends on the concentration of the thiosulphate.

Questions MS1–MS4 are concerned with a study of this reaction.

A pair of pupils made up 20 cm³ portions of each reactant at a known concentration and temperature. The 20 cm³ portions were kept in boiling tubes, and reaction was begun by pouring a pair of reactant solutions into a small flask. The mixture was observed, and the time taken for a faint but visible precipitate of sulphur to appear was measured.

The pupils' results are given in the table.

	Temperature (°C)	Composition of 20-cm³ portion of HCl		Composition of 20-cm³ portion of thiosulphate		Time (seconds)
		(cm³ 2 molar HCl)	(cm³ water)	(cm³ 0.1 molar thiosulphate)	(cm³ water)	
1.	19	20	0	20	0	39
2.	19	20	0	15	5	52
3.	19	20	0	10	10	78
4.	19	20	0	5	15	157
5.	19	15	5	20	0	39
6.	19	10	10	20	0	39
7.	19	7.5	12.5	20	0	40
8.	19	5	15	20	0	42
9.	19	2.5	17.5	20	0	50
10.	19	1.25	18.75	20	0	95
11.	6	20	0	20	0	82
12.	15	20	0	20	0	48
13.	25	20	0	20	0	29
14.	31	20	0	20	0	20
15.	42	20	0	20	0	14
16.	49	20	0	20	0	10
17.	60	20	0	20	0	6

When the precipitate of sulphur first appears the extent of the reaction is small. If it is assumed that the sulphur can always be first seen when the same small amount of sulphur has been formed, the time taken up to this point will give a measure of the initial rate of the reaction.

Which of the following will be the relation between the time determined by the pupils and the initial rate of the reaction?

(a) Time = initial rate
(b) 1/Time = initial rate
(c) Time ∝ initial rate
(d) Time ∝ (initial rate)²
(e) 1/Time ∝ initial rate
(f) Time² ∝ initial rate

MS2. This question is concerned with the influence the thiosulphate concentration has on the initial rate.

(a) Which of the readings in the table of Question MS1 would you use to determine how the rate of the reaction is influenced by the thiosulphate concentration?

(b) Use these readings, perhaps drawing a graph, to find the connection between the rate of the reaction and the thiosulphate concentration. Explain clearly how you have arrived at your conclusion.

(c) Estimate the time taken for the initial precipitate of sulphur to appear on mixing 10 cm³ of 2 molar HCl and 10 cm³ of 0.1 molar thiosulphate.

MS3. This question is concerned with the influence the HCl concentration has on the initial rate.

(a) Which of the readings in the table of Question MS1 would you use to determine how the rate of the reaction is influenced by the HCl concentration?

(b) What conclusion can you draw about the connection between rate and acid concentration?

(c) You are provided with 2 molar thiosulphate and 0.1 molar HCl, and are using them at about these concentrations to find how the rate depends on the thiosulphate concentration (by varying the thiosulphate concentration while keeping the acid concentration constant). What result would you expect to obtain?

MS4. This question is concerned with the influence of temperature on the initial rate.

(a) Plot a graph of initial rate of reaction (y axis) against temperature.

(b) Can you account for the way in which the rate changes with temperature?

(c) Estimate the time taken for the precipitate of sulphur to appear at 0 °C.

(d) Is it possible to estimate the time taken for the precipitate to appear at 100 °C?

(e) Can you find any functions of rate and temperature (in K, which is °C + 273) which when plotted graphically give a reasonably straight line? Does, for example, a plot of log rate against T give a straight-line plot?

MS5. The next three questions are concerned with the relation between particle size and surface area, with particular reference to surface catalysts.

(a) The mass of a 1 cm cube of metal is 21.4 g. What is the surface area of the cube?

(b) What is the volume of this cube of metal?

(c) How many small cubes can be obtained from 21.4 g of this metal if the length of the sides of the small cubes is to be 0.2 cm?

(d) What would be the total surface area of 21.4 g of these 0.2 cm cubes?

(e) The metal is found to be a surface catalyst for several reactions. What can you say about the rate of these reactions when (i) no catalyst is present, (ii) a 1 cm cube of metal is present, (iii) 21.4 g of 0.2 cm cubes is present, (iv) 1 milligramme of 0.2 cm cubes is present?

(f) What is the metal?

MS6. This question is about the same metal as in Question MS5.

(a) What would be the surface area of 21.4 g of 0.002 cm cubes?

(b) What length of wire (cylindrical) of diameter 0.002 cm could be drawn from 21.4 g of the metal?

(c) What would be the surface area of 21.4 g of this wire?

(d) What length of wire of diameter 0.000 02 cm could be drawn from 21.4 g of the metal?

(e) What would be the surface area of 21.4 g of this wire?

(f) Which of the following numbers will be closest to the diameter of the wire in 'number of metal atoms' when the diameter is 0.000 02 cm: (a) 1, (b) 10, (c) 100, (d) 1000, (e) 10 000?

MS7. 1 g of a metal is available to act as a surface catalyst for a reaction. Arrange the forms of metal given below in order of probable catalytic effectiveness, with most effective first:

(a) One cube.

(b) A number of cubes of side 0.002 cm.

(c) A number of cubes of side 0.000 02 cm.

(d) A length of wire of diameter 0.002 cm.

(e) A length of wire of diameter 0.000 02 cm.

MS8. This question is concerned with the number of collisions that molecules must undergo before they will react.

A study has been made of the decomposition of ozone (trioxygen) molecules by a silver(I) oxide surface of known area. The reaction is

$$2O_3 \rightarrow 3O_2$$

In the experiment, a stream of ozone was passed at different speeds over the silver(I) oxide in a tube. A determination was made of the ozone which passed through without being decomposed, and this was related to the average number of times the ozone molecules had hit the catalyst surface. Some of the results are shown in the table below.

Percentage of ozone molecules passing through without being destroyed	Average number of times an ozone molecule hits the catalyst surface
84	0.2
71	0.4
59	0.6
48	0.8
36	1.0

(a) Plot the results on graph paper, with percentage of ozone as the y axis.

(b) Estimate the average number of collisions required for reaction to occur.

(c) What would be the percentage of ozone passing through if the catalyst was removed?

(d) Why is the answer to (c) unexpected, and how could it be explained?

(e) Would you expect the answer to (b) to be a whole number?

MS9. The curves given below show the energy distribution of the ozone molecules used in the experiment of Question MS8. The shaded area under each curve is meant to represent those molecules which decompose on their first impact with the catalyst surface. The percentage given by each curve indicates how much of the total area under the curve is shaded.

Which ONE of the shaded areas is most likely to be correct?

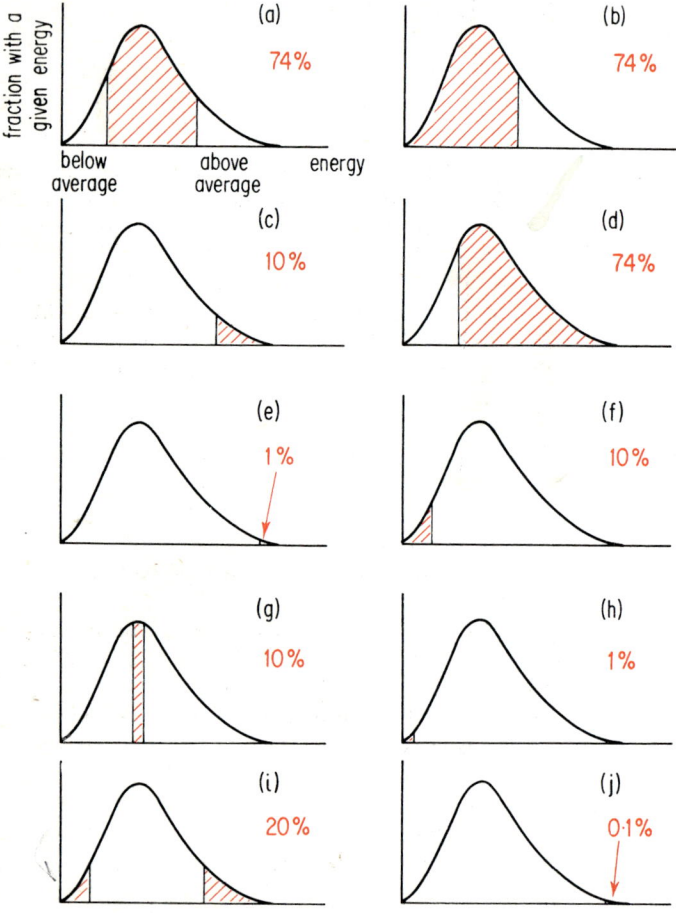

MS10. The reaction of oxygen with a heated metal filament has been studied, and it has been found that only 1 molecule of

oxygen in every 1000 which strike the metal surface actually reacts. At higher temperatures the number of molecules reacting with the surface increases.

(a) What happens to the other 999 in every 1000?

(b) If the energy distribution of the oxygen molecules is that given in (a) to (j) of Question MS9, which one of the shaded areas is most likely to represent the oxygen molecules which react with the catalyst surface?

MS11. In an investigation of the condensation of water vapour on to a liquid water surface it was found that only 1 molecule of water vapour in every 100 which hit the surface actually condensed.

(a) What happened to the other 99?

(b) Suggest methods by which the figure might have been determined.

(c) If the energy distribution of the molecules of water vapour is that given in (a) to (j) of Question MS9, which one of the shaded areas is most likely to represent water molecules which can condense?

MS12. The reaction of oxygen with a heated metal filament has been mentioned in an earlier question. At 1000 °C it was found that only 1 molecule in every 1000 reacted on striking the metal surface. At 2500 °C it was found that 150 molecules in every 1000 reacted on striking the surface.

The increase in the average energy of oxygen molecules for a temperature change from 1000 °C to 2500 °C is about 50 per cent. Can you account for the enormous increase in the rate of reaction resulting from a relatively small increase in the average energy of the molecules?

MS13. A pupil was interested in investigating the rate at which iodine goes into solution in water, and also the equilibrium position when the water was saturated with iodine. The next eight questions are concerned with this investigation.

The pupil took a large volume of water at 17 °C in a flask. At the beginning of the experiment he tipped in about 5 g of iodine, corked the flask and began to shake it vigorously. Every two minutes he uncorked the flask for a moment, poured out 20 cm³ of the solution into a measuring cylinder, corked the flask and began shaking it again.

While he continued to shake the flask, his partner estimated the concentration of the iodine in the 20 cm³ sample. [He did this by titrating against 0.002 molar sodium thiosulphate, using starch solution to indicate when sufficient thiosulphate had been added to react with the iodine according to the equation.

$$2Na_2S_2O_3 + I_2 \rightarrow 2NaI + Na_2S_4O_6]$$

Here are the results of the experiment:

Time after start (minutes)	Concentration of iodine (moles/litre)
2	0.000 20
4	0.000 40
6	0.000 58
8	0.000 70
10	0.000 80
12	0.000 88
~~14~~	0.000 92
20	0.001 00

There was no change in the iodine concentration after twenty minutes. Excess solid iodine remains.

Plot a graph of concentration of iodine (y axis) against time and use this to estimate the iodine concentration after one, five and fifteen minutes.

MS14. At which ONE of the following times in minutes from the start did the iodine appear to be dissolving at the greatest rate: (a) 3, (b) 5, (c) 7, (d) 9, (e) 11, (f) 19?

MS15. Arrange the times given in Question MS14 in order of apparent rate of dissolving, with highest rate first.

MS16. Which ONE of these values is the apparent rate of dissolving at the start, expressed in moles of iodine/litre/minute: (a) 0.000 05, (b) 0.0001, (c) 0.0004, (d) 0.0127, (e) 0.0254?

MS17. Five possible curves are shown below. Also included is the original curve obtained by the pupil.
If the pupil had repeated the experiment of Question MS13 with 10 g of iodine at 17 °C, his curve would probably have been close to one of the following. Which one is it: (a) 1, (b) 2, (c) 3, (d) 4, (e) 5?

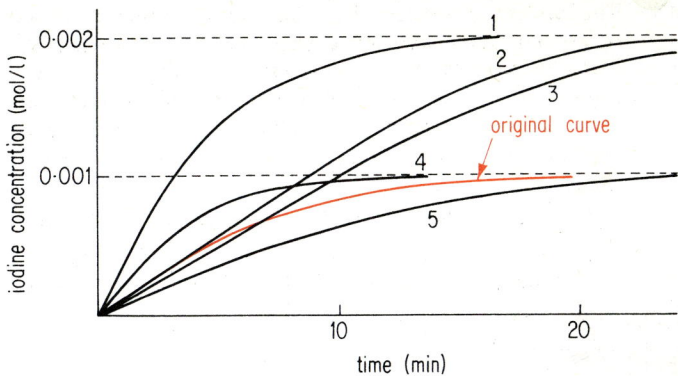

MS18. If the pupil had repeated the experiment of Question MS13 with 5 g of iodine in a finer state of division, his curve would probably have been close to (a) 1, (b) 2, (c) 3, (d) 4, (e) 5.

MS19. If the pupil had repeated the experiment in Question MS13 with the original 5 g of iodine, using water at 38 °C, the curve is most likely to have been closest to (a) 1, (b) 2, (c) 3, (d) 4, (e) 5.

MS20. Which one of the following explanations best fits the observation that iodine becomes more soluble on raising the temperature?

(a) The iodine will dissolve more rapidly and therefore at equilibrium there will be a higher concentration of iodine.

(b) The iodine will dissolve more rapidly and precipitate less rapidly, so that at equilibrium there will be a higher concentration of iodine.

(c) The iodine will dissolve more rapidly and precipitate more rapidly, so that in fact the iodine concentration should stay the same.

(d) The rate of both dissolving and precipitating will increase with temperature rise, but in this case the increase in the first is greater than the increase in the second.

(e) The iodine molecules once in solution will have a greater energy at a higher temperature and will therefore be less likely to return to the solid state.

47

MS21. When excess calcium carbonate is heated in a closed and previously evacuated container the pressure inside the container rises as the following equilibrium is established:

$$CaCo_3(s) \rightleftharpoons CaO(s) + CO_2(g)$$

The gas pressure gives a measure of the extent of reaction. It is found that at a steady temperature the pressure does not change with time or on addition of more calcium carbonate.

For which one of the following reasons is the pressure constant at a steady temperature?

(a) If the temperature of a fixed mass of gas is fixed, the pressure will be fixed.

(b) When equilibrium has been reached, no more gas is produced.

(c) At equilibrium the gas is being produced at the same rate as it is going back to the solid.

(d) When equilibrium has been reached, no more gas returns to the solid.

(e) At equilibrium there is no more calcium carbonate, so no more gas is formed.

(f) Only certain molecules of calcium carbonate are of a type which can decompose, and these have decomposed.

MS22. Here is some information about the carbon dioxide pressure over calcium carbonate at various temperatures.

Temperature (°C)	Pressure (mm mercury)
691	19
727	44
749	72
786	134
800	183
819	235
857	420
881	603
907	875

Draw a graph of temperature against pressure (y axis), and predict the temperature at which the pressure of carbon dioxide will be 1 atmosphere (760 mm Hg).

MS23. This question is concerned with the dissociation of calcium carbonate on heating, quantitative information for which is given in Question MS22.

(a) Does the curve obtained in Question MS22 resemble that obtained in Question MS4(a)? Can you put forward any explanation for the way in which the carbon dioxide pressure changes with temperature?

(b) Is there sufficient information to predict (i) the temperature at which the pressure will be 10 atmospheres (ii) the carbon dioxide pressure at room temperature?

(c) The temperature of a roaring bunsen flame is supposed to be over 1000 °C. At this temperature the carbon dioxide pressure should be several atmospheres, and yet it is difficult to bring about appreciable decomposition of calcium carbonate by heating in a test-tube. Can you put forward any explanation for this?

(d) A pupil believes that he has achieved complete decomposition of calcium carbonate by heating in a test-tube. Suggest methods by which he might determine whether this is correct.

MS24. Here is some information about the vapour pressure of tungsten at various temperatures.

Temperature (°C)	Vapour Pressure (mm Hg)
3016	0.001
3309	0.01
3900	1
4507	10
5168	100

(a) Can you give a simple explanation for the way in which the vapour pressure changes with temperature?

(b) Can you use these figures to predict the boiling point of tungsten?

(c) Can you discover in a reference book the working temperature of a tungsten lamp? Do you think that the vaporization of the wire is significant at this temperature?

(d) Can you find out the function of the quartz and the iodine in a quartz-iodine lamp?

(e) How would you set about attempting to boil tungsten in the laboratory?

Appendix

Answers to some of the 'thinking' questions

2 (*a*) A black film of copper(II) oxide is formed on the surface.

(*b*) The red lead decomposes into lead(II) oxide (PbO) and oxygen.

$$2Pb_3O_4 \rightarrow 6PbO + O_2$$

Both these changes are irreversible in the conditions mentioned.

2 On gentle heating, roll sulphur melts to a pale yellow, mobile liquid. If this liquid cools it solidifies, reforming solid sulphur.

When iodine crystals are heated under ordinary conditions they change directly to purple iodine vapour, without passing through a liquid stage. When iodine vapour cools it returns directly to the solid phase and deposits small crystals of iodine. This change is known as sublimation.

2 The solid becomes warm to the touch.

3 It will be necessary to obtain a sample of calcium oxide which does not effervesce on the addition of hydrochloric acid, thus proving the absence of any calcium carbonate. To obtain such a sample it may be necessary to heat some stock calcium oxide very strongly and then allow it to cool in a desiccator.

3 The hydrogen must be thoroughly dried and the iron(II) diiron(III) oxide must be free from iron. There's an added difficulty in that some samples of iron oxide are magnetic. Show that a sample of the oxide does not effervesce with dilute hydrochloric acid.

4 The equilibrist will restore his balance by moving his pole up and down at right angles to the wire.

5 No. He would remain in the same place, relative to the two floors of the tube station.

6 We start with all the balls in A and both air supplies blowing. Balls begin to pass into B and slowly the number increases until we get the impression that a position is reached when the numbers in A and B are roughly equal.

If we introduce a few red balls we can follow their paths.

49

6 We often see them travelling from one compartment to the other and sometimes back again.

11 Yes they could. This would be a chance factor, depending on the numbers of balls involved.

If released at C the ball would come to rest at D.

If released at A the ball would end up at G.

B is the lowest release point to allow the ball to pass over the hump.

11 The vertical height EF might be considered to correspond to the activation energy.

15 The lower the barrier height the easier it is for the balls to pass over.

A millisecond is one thousandth of a second

A microsecond is one thousandth of a millisecond

16 A nanosecond is one thousandth of a microsecond.

16 The rate is now zero; the reaction has ceased.

Carry out a suitable test for hydrochloric acid, or rather for its absence, but not a test for the chloride ion, for calcium chloride is also present. Test with a piece of magnesium ribbon, after boiling off dissolved carbon dioxide.

17 With crushed marble the reaction would be faster.

17 With 1 M acid the reaction would be slower than with 2 M acid. Graph A represents the reaction with the more concentrated acid.

18 In both gaps write 'increase'.

18 If you take 40 cm^3 of stock solution and add 10 cm^3 of water and shake, then the resulting solution will be 4/5 of the concentration of the original. Similarly 30 cm^3 of

stock solution and 20 cm^3 of water will give a concentration which is 3/5 that of the original.

18 You can consider, say, 1 cm^3 of the stock sodium thiosulphate solution as a unit of concentration. If then you use 50 cm^3 of stock solution you have 50 units of concentration. If you take 30 cm^3 diluted with 20 cm^3 of water, then this represents a concentration of 30 units.

20 Both tubes might be buried to the same depth in a tin of sand and then the tin could be heated centrally with a burner.

Oxygen was given off from both tubes, but first from the tube containing the copper(II) oxide.

If you heat copper(II) oxide alone, no oxygen is evolved.

21 As well as factors such as concentration of the peroxide and its temperature, you might also investigate the effect of different catalysts and of different quantities of any one catalyst.

22 The surface area of the powder is much greater, and hence the reaction rate will be higher under corresponding conditions.

24 The path from A to B through the tunnel would require less energy. (But there might be a toll to pay!)

29 In theory it would never be possible to transfer all the iodine from the trichloromethane layer, but in practice after a great number of treatments the quantity left would be so small as to be almost invisible.

29 To show that when silver is added to an iron(III) solution some iron(II) ions are formed, test with potassium hexacyanoferrate(III), when a deep blue colour will be formed.